SAVING THE PRAIRIES

SAVING

Ronald C. Tobey

The PRAIRIES

The Life Cycle of the Founding School
of American Plant Ecology, 1895-1955

UNIVERSITY OF CALIFORNIA PRESS   Berkeley   Los Angeles   London

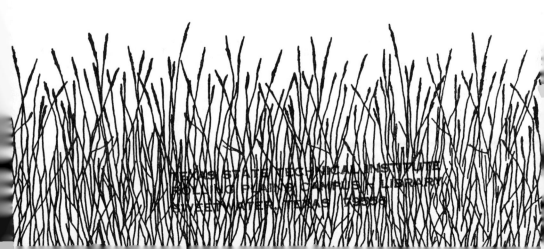

University of California Press
Berkeley and Los Angeles, California

University of California Press, Ltd.
London, England

Library of Congress Cataloging in Publication Data

Tobey, Ronald C.
  Saving the prairies.

  Bibliography: p. 285
  Includes index.
  1. Botany—Ecology—History.   2. Botany—United   States—History.
    3. Prairie ecology—United States—History.   I. Title.   II. Title:
American plant ecology, 1895–1955.
QK901.T6   581.5′09   80-28200
ISBN 0-520-04352-9

Printed in the United States of America

1 2 3 4 5 6 7 8 9

To Amy and Liz

CONTENTS

ACKNOWLEDGMENTS

I am grateful to helpful friends who contributed their advice, criticism, and encouragement during the past six years. Pamela Smith has been my programmer and the hidden collaborator on my quantitative analyses. Her training in sociology, her unflagging ingenuity in manipulating our computers, and her intelligence provided many of the moments of pleasure in this study. John Phillips has been a steady guide through the fits and starts of my quantitative analysis of career patterns and generations of literature. Charles Taylor frequently shared my ideas about the life cycle of these scientists, contributing the expertise of his training in population biology. John Moore—bless you, my dear friend—startled me with more embarrassing questions and obscure references to magna opera in my field than I care, indeed, dare, remember; that he thought well of me and of some of my ideas continually encouraged me. Joe Svoboda, archivist of the University of Nebraska, went out of his way to aid my search for manuscripts and information. I simply could not have accomplished the archival research without his kindnesses.

Frank Egerton and Robert McIntosh read one version of this manuscript with great attention. I am indebted to their suggestions for the revision of several theses as well as correction of detail.

Frank Vasek, my colleague in the Department of Plant Sciences, scrutinized the final version of the manuscript, modernizing some botanical terminology and pointing my language toward contemporary nuances of botanical meaning.

Versions of chapter 5, "The Life Cycle of Grassland Ecology," were presented at three different seminars. At the International Conference on Quantification in the History of Science, summer, 1976, Derek Price, Jerry Gaston, and Barbara Rosenkrantz made pertinent comments that clarified my sociological analyses. A special presentation of the chapter was made to the University of California, Berkeley, Ecology Seminar, fall, 1976, organized by Arnold Schultz. The discussion of my interpretation by scientists was lively, and while I may not have entirely convinced them of all of my theses, their questions have helped me to frame a more persuasive presentation of my work. Finally, the chapter was presented to my friends in the biology department at my own campus, in the winter of 1978. Once they recovered from the shock of seeing a historian use graphs, they had challenging questions, particularly on the influence of H. A. Gleason; such comments, as in my other conversations with scientists, were helpful to me in assessing my own views.

My ideas on the ideology of ecology in chapter 7 were presented at the annual meeting of the Society for the Social Study of Science, fall, 1977. Diana Crane, whose work on invisible colleges was used in chapter 5, was gracious, encouraging, and helpful then, as she has been in correspondence.

On other occasions, other colleagues have generously contributed to my work: George Gillette, William Thompson, Bob Hine, Ed Gaustad, Leon Campbell, Kenneth Barkin, Irwin Wall, Gene Gressley, Robert Bessey, Ted Pfeifer, David Crosson, J. Max Hoffman, Irene Brown.

I have been fortunate to have had conscientious and interested student research assistants. Particularly, I thank Henry Lowood, Ann Eden, Yaya Fanusie, Phyllis Okeneske, William McClosky, and Dorothy Mandal.

Gabriele Gonder helped me through some nearly impenetrable German language botany.

The typing of the manuscript versions of this book has been done efficiently and cheerfully by Kathleen Spore.

A portion of the statistical analysis and text preparation was conducted in the Laboratory for Historical Research at the University of California, Riverside. I appreciate the aid of Dr. Charles Wetherell, Director.

The research travel for this study has been supported by a University of California fellowship, which allowed me to examine the Bessey and Clements collections, and a National Endowment for the Humanities fellowship, which permitted me to read in the British Museum and Kew Library in nineteenth-century botanical literature. The University of California Academic Senate Research Committee has faithfully supported the computer work, my programmer, and my research assistants with research grants. Finally, the dean of the College of Humanities and Social Sciences, University of California, Riverside, provided the funds for the preparation of this manuscript.

INTRODUCTION

These are the gardens of the desert, these
The unshorn fields, boundless and beautiful,
For which the speech of England has no name—
The Prairies. I behold them for the first,
And my heart swells, while the dilated sight
Takes in the encircling vastness.
                              —William Cullen Bryant

Popular writers today assume that "ecology" sprang from the tran-
scendental naturalism of Emerson and Thoreau, or from the preservation movement represented by John Muir and John Burroughs.
Historians know this is not true. While the middle-class tradition of
nature enthusiasm has been since the 1880s a background to the rise
of scientific ecology, the key insights making possible the scientific
paradigm of the first generation of ecological research were derived
from utilitarian and scientific problems, not from the diffuse affecta-
tions of an urban public attached to the nature of summer camp.
Unfortunately, we do not know much about this rise of a scientific
specialty that has in the past ten years shared media headlines with
important political and social events. We have only a pitifully small
journal literature and only one general narrative account of the
history of ecological ideas. For two generations, ecologists have been
inextricably involved in public issues of the preservation and utilization of our biological resources. Their knowledge and their values
have been crucial to the forest reserve system, game management,
agricultural exploitation of the federal lands, and disaster relief.
During the hectic decade of the 1960s, they supplied leaders in the
diverse coalition of preservation and planning interests that blossomed in "Earth Day," and other more substantial political movements. Following the passage of the National Environmental Policy
Act in 1969, they have probably been closer to local public issues
than any other group of scientists, as "impact reports" have repeatedly forced the public to face the hard choices in planning our

1

environment. Even where others—archaeologists, natural resource economists, historians—have played a more prominent role than ecologists in environmental management, frequently, perhaps inevitably, their language has reached metaphorically to scientific ecology, to the relatedness of things. It is in the public interest to know more about the values of these scientists.

Urgent reasons in the professional history of science similarly press us for a study of ecology. Since the early 1960s, a growing contingent of historians and philosophers of science has been convinced that science is not the objective, nonpersonal, nonideological activity the popularizers of science have frequently led the public to believe it is. This contingent has grown so large within the profession—are we now a majority among our colleagues?—that we can be surprised to discover that the general educated public does not share our new understanding of science. The specialty of ecology therefore especially cries out for examination from the new professional point of view. Scientific knowledge reflects the character of a nonscientific set of values; doesn't this imply that the scientific study of environmental impact is one set of values judging other sets of values?[1]

The story of the grassland ecologists is intrinsically interesting for the drama of their struggle to understand and to preserve one of the great biological regions of the world, and because this struggle created the science of ecology, considered in a professional sense, in the United States. By the 1930s, their theory of succession provided the dominant framework for all ecological research, animal as well as plant, in America and Great Britain, and some other parts of the world, such as South Africa and Australia. Their community was tightly knit by graduate training at a few important universities, by joint research and coauthored publications, and by their interwoven careers in the research complex of the Midwest, namely the land-grant universities, experiment stations, and the research agencies of the federal government. But most of them were bound to each other by more than these sociological ties. At the core of their community was the shared experience of the prairies and plains. The first generation—C. E. Bessey, Roscoe Pound, Frederic Clements—was raised on the frontier and entered botany just as the successive booms of settlement were breaking upon the virgin soil. The next generation of John E. Weaver and many secondary contributors to the specialty, such as Paul Sears, was still able to experience isolated fragments of pristine prairie and to feel the entire presence of the original plains as a deepening echo in their lives. Sears, who moved to Yale University for a distinguished career in pollen analysis and forest ecology, reminisced about a childhood railroad journey out of the Great Lakes forests onto the prairie, "a vast gray-green sheet dappled and dancing in a light breeze. The impression of it, after seventy years, remains in the realm of sensation, not to be translated adequately into words."[2] So David Costello, a prominent federal grassland ecologist, began a popular book on the prairie by remembering the primary fact of his childhood: "Years ago, when I was a boy and the prairie world was still there."[3] The most lyrical scientific invocation of the prairies was uttered, perhaps rightfully, by John Weaver, the leading scientist of the second generation. In 1944, he and his students had recently concluded a struggle to preserve the prairie against the drought from 1933 through 1941. The struggle had not been entirely successful. Vast tracts of tall grass had died. A half-century of cultivation, of turned sod, of wind erosion, and spreading bluegrass had, perhaps permanently, suspended the laws of vegetational development, so that the original prairie would not return without man's protection. Weaver chanted the seasonal cycles of the prairie flora. The spring on the prairie brings "the yellow of the golden parsnip, the bright pink or purple of the prairie phlox, the white masses of flowers of New Jersey tea, and the buffalo bean with its abundance of violet-purple flowers." The summer aspect of the prairie wears "white and purple prairie clover, black-eyed Susan, tick trefoil, wild licorice, wild bergamot, and rosinweeds."

The central fact about the prairie was its natural stability and tough perseverance. Grassland-dominated communities were so perfectly constructed, after thousands of years of evolution, that weeds could not penetrate, even when the land was settled and grazed. As long as the prairie vegetation was not destroyed and replaced by man, he wrote, with important qualification, the prairie would remain: "It is a slowly evolved, highly complex organic entity, centuries old. It approaches the eternal."[4]

The prairies were the heart, the enduring strength of the American continent. John Weaver made no reference to the world war destroying Europe and Asia; but it was clear in his little article that he believed that as long as Americans preserved and lived in harmony with the great midcontinent grassland, the symbol of its democratic community, that they too would endure. The problem was that agriculture had not been "in harmony" with the grassland environment. Weaver felt personally fortunate to have lived in a section of Nebraska where the original prairie encountered by nineteenth-century pioneers still existed, its grasses not grazed down by cattle, its soil not ruptured by a plow. He had lived where the prairie had " 'resisted civilization' the longest."[5]

Even the ecologists themselves recognized that the power of their shared experience of the grassland and the tacit bonds of their community raised a significant scientific problem. How could scientists critically test assumptions so central to their experience that it would not occur to them the assumptions were open to challenge? Or how could they test scientific assumptions so central to their experience that they rarely reached articulation? How could scientists with similar training and research experience obtain sufficient distance from themselves to test realistically their fundamental ideas? Clements and Weaver succinctly stated the problem in *Experimental Vegetation* (1924):

> The development of the views as to the nature and structure of the grasslands of North America illustrates the need of objective methods of determining vegetation units and their relationships. This is all the more convincing, since the ecological investigation of the prairie and plains has been the work of a group with the same general training and outlook.[6]

So stated, Clements's and Weaver's problem of objectivity was identical with the problem of objectivity for any group making claims to the universality of its epistemology and the validity of its values— *any group*, whether political, social, or scientific. Like most scientists, they believed they could invent methods of investigation and proof to establish "objectively" their assertions about the natural world, assertions that were independent of the social role of science. In this book, we will see that this quest was not successful.

I would be happy if I could say that this book is the big, definitive work on the history of scientific ecology in America that not only gives us the general narrative we need but also gives us an analysis of ecology's specialized knowledge from the postpositivist point of view, relates the scientists to our political history and to their lay supporters and critics, and does all this for all sorts of ecologists. I would also be terminally exhausted. For such a history is beyond the capacity of the historical profession at the moment. We must settle for some preliminary, fragmentary histories now, with a grand synthesis later. I have chosen to provide the history of the first coherent group of ecologists in the United States, the grassland ecologists of the Midwest. I have treated this group as a case study, for their history touches most of the major themes that concern us about the role of ecologists, and generally about the role of scientific expertise, in our society. The grassland ecologists' history also touches significant issues in current professional debates over the character of science. These issues I take up in their appropriate places in the narrative.

I have tried to show that the content of the grassland community's knowledge was intimately related to the social structure of the community and to the role the community played in American society. The grassland scientists believed, at the outset of their history in the 1890s, that vegetational change was inevitably progressive and over-rode man's intervention in the environment. They came to believe, in later years of great biological and economic crisis, that vegetational change was not necessarily progressive or self-repairing, and that repair might occur only as a result of man's management of the environment. This fundamental shift in their profoundest assumptions was tied to their role in our society. That this shift was analogous to the movement of American history from nineteenth-century innocence to twentieth-century culpability should be one more clue to the close relationship between scientific knowledge and the large society in which it is situated.

Lest any reader naively jump to the conclusion that there is a sociological monograph hidden inside this historical essay, let me demur. The professional literature of the sociology of science is enormous, sociologists' theories about science are many, and I have not tried to test systematically the validity of any of them. I have used those theories that seemed to me relevant to understand the social features of the community of grassland ecologists. And I have tried to use them with technical accuracy, rather than simply as jargon. Sociologists may complain that this is a free ride, unfair and method-ologically unsound; to which I can only reply, nonmalicious trespass. Without a basic narrative history of the founding school of American plant ecology, it would be impossible to test meaningfully any sociological theories about that history. I hope that this effort approaches the oft-cited ideal of sociologically informed, but not martyred, history.

If my book is not a straightforward sociological treatise, it is also not a straightforward history of ideas. Readers expecting the conventional narrative of intellectual history will be confused. I have not simply isolated the major ideas of the Clementsian ecological theory and followed their development in Clements's published writings, prefaced by a reconstruction of their precursors and suffixed by their denouement in the hands of his critics. This style of intellectual history is in retreat in the profession by reason of its embarrassing methodological fallacies. At a recent conference, for instance, one autopsy on intellectual history diagnosed the pathology in terms of unwarranted generalization beyond the evidence, unrepresentativeness of the ideas under examination, and a fundamental confusion about the causal relations between material and social structures and ideas.[7] To avoid these debilitating flaws in the history of ideas, I examine the history of Clementsian theory within the framework of a case study of a microparadigm, in terms of Thomas Kuhn's theory of scientific change, as modified recently by sociologists of science. Kuhnian theory determines the structure of this case study and, hence, of this book. My interest is in the formation and dissolution of the microparadigm of grassland ecology, not of the development of its leading principles or of their extension to new examples during the period of normal science. My generalizations about the extent to which Clements's ideas were held and used are bounded carefully by the grassland community as the study group. The representativeness of views is empirically tied to citation counts and qualified by traditional literary analyses. And the relationship between ideas and the social and material structures is examined in terms of modified Kuhnian sociological theory. While this program may be modest and unsatisfying to readers accustomed to the intellectual vistas of overgeneralization, it should satisfy readers who are stimulated by the bracing wind of rigor and procedure.

The case study begins with the immediate social and intellectual setting within which Frederic Clements's ideas were born, the study of botany under Charles Edwin Bessey at the University of Nebraska in the 1880s and 1890s. Then I turn to the larger environments of Bessey's botanical study—the great explosion of laboratory instruction and research in American universities following the Civil War and the coincident upsurge of public interest in nature study and the wild. These settings contributed important features to the Clementsian microparadigm, such as mathematical rigor, experiment, and utility. I also argue that these settings channeled certain ideas into Clements's theory and screened out others. In particular, they brought to Clements and Roscoe Pound the European tradition of plant geography associated with Oscar Drude and cast as less important another tradition of plant geography, that of Eugene Warming. Furthermore, they provided Clements with a perspective from which to view the intellectual systems and philosophies coming to him from the nineteenth century which bore some resemblance to the views he later espoused. For instance, the philosophies of Plato, Hegel, and Herbert Spencer all emphasized the reality of organic collectivities and their ontological development. Clements undoubtedly read some works of these philosophers; does this mean that his own principles, which appear similar, were "influenced" by theirs, or were the "result" of having read them? Not necessarily. The immediate "Bessey system" in which Clements was taught was hostile to the character of those philosophies. Rather, I argue that Clements picked up two major streams of thought. One was Oscar Drude's plant geography, which culminated a tradition of European floristics originating with Alexander von Humboldt. The other stream was the tradition of Spencerian sociology, especially as expressed by Lester Frank Ward, whose liberalism coincided with the pragmatism and scientism of Bessey's land-grant philosophy.

The establishment of the Clementsian theory occurred in a sociological, as well as an intellectual, competition with an opposing theory, that of the University of Chicago's great botanist, H. C. Cowles. I analyze this competition and the subsequent victory of the Clementsian theory largely in sociological terms as the victory of a school of scientists. Establishment of scientific ideas here appears less as the victory of truth over error than the building of networks of collaboration, placement of graduate students in strategic jobs, and who cites whom. Victory does not necessarily go to the biggest legions, but the cliché strikes one as apt. My emphasis on sociology may distress some historians and scientists, whose experience has been with normal science and who believe that intellectual victory in science should be described in terms of the accumulation of empirical facts and the verification and falsification of hypotheses. I have not narrated the contents of the grassland ecological literature of the normal science period, and I have not narrated each modification of Clements's ideas by his many disciples, because these were not central to the establishment of Clements's theory. To the contrary, they assumed its establishment and occurred within the framework of commitment to the theory; they were not efforts to test or falsify the theory. These sections on the Nebraska school heavily rely on statistics for illustration, summarization, and analysis. I hope for the indulgence here of the nonprofessional reader. These statistical discussions are innovative in the history of science and central to the case study approach. Readers not interested in these efforts to advance professional analysis should turn to the concluding chapters.

Finally, I describe the breakup of the Clementsian microparadigm. This destruction occurred when philosophical shifts drove away secondary adherents and a great natural phenomenon—the midwestern drought of the 1930s—challenged primary advocates. The English defender of Clementsianism, A. G. Tansley, provides us with a view of the process by which a friend of the theory, who was not party to the Bessey system, gradually lost his philosophical toleration of the theory. At the same time that political and social changes in the 1930s in Great Britain made Clementsian principles untenable for Tansley, the environmental changes in the grasslands in the 1930s, coupled with the political and social changes of the New Deal era, broke the Clementsian faith of grassland scientists in the United States.

1
MISSIONARIES FOR BOTANY:
C. E. Bessey and His Students

> Dr. Bessey, . . . Perhaps you remember telling your students that "if they ever went to any country High School, to act as missionaries in the interest of botany." So I have tried to comply with this earnest request.

A Leading Department of Botany

Charles Edwin Bessey was a great teacher. He transformed the dry, girls' school subject of basic botany into a gospel course for the scientific, laboratory method. He transformed raw youths from the new farmlands of booming Nebraska into missionaries for science. "The mission of the true teacher is to burn the thoughts in, brand his pupils for life," one of his disciples wrote, repeating his own words back to him.[1] Bessey must have been pleased when students testified to his influence during their brief college days. When a "farmer's wife," living with satisfaction a country life, sent a water-moss to her former teacher for identification, she expressed her gratitude for "small kindnesses, long forgotten, which gave a scientific bent to my inherited love of flowers and so made a life fuller and happier."[2] In little more than two brief decades, from 1884 to 1907, while he was in his prime and before the aggressive spirit of the University of Nebraska was reined in by political attack, Bessey created one of the nation's leading departments of botany. He fused together a scholarly community that included precollege students in "prep bot" and a famous succession of doctoral scientists. In a setting not only of diverse interests, from teaching elementary botany to controlling devastation of cultivated grasslands, but also of varied personalities—one doctoral scientist held down a law career and another was inspired by the classics—C. E. Bessey benevolently but critically nurtured into life the new science of ecology.

Before he came to the University of Nebraska, Bessey had established himself as a teacher, scientist, and administrator at the state university of Iowa. He was, indeed, a product of the new state university system that was created by the Land Grant Act. While at

9

Nebraska he contributed to the continued reformation of that important American educational innovation by helping write the Hatch Act of 1887, establishing the agricultural experiment stations. He was born in 1845 on a farm in Milton, Ohio, and educated after the Civil War at Michigan Agricultural College. He took his bachelor of science in 1869 and trekked west for his first teaching job at Iowa State College of Agriculture at Ames. He continued his studies while teaching and received his master of science from Michigan in 1872. Then he took a year off to study under the elderly Asa Gray, presiding spirit of American botany at Harvard. That year, together with another brief sojourn with Gray in 1875–76, provided Bessey's only educational experiences outside the western state-supported universities. He never studied in Europe; nevertheless, it was by this deeply American scientist at his distant scientific outpost in Iowa that the first American textbook of botany was written which incorporated the latest advances of European experimental and microscopic botany. *Botany for High Schools and Colleges* (1880) made Bessey a famous man. Writing it coincided with his doctorate, which he took from Iowa State in 1879. Bessey was no less an administrator than he was scientist and teacher. In 1882 he was acting president of Iowa State and from 1883 to 1884, vice-president. It was, ironically, his reputation as administrator that brought an offer to go north to the University of Nebraska, where his greatest accomplishments as a scientist were to be made.[3]

He arrived in Lincoln, Nebraska, in the fall of 1884, with triple duties. He was dean of the Industrial College, professor of botany and horticulture, and Nebraska state botanist. The Industrial College, founded in 1877, had been under criticism for low enrollment and for being a trade school. The regents of the university expected Bessey to lay a new academic foundation for the college and to transform its farm into a genuine experimental facility. By 1887, he had accomplished these objectives.[4]

As the university's botanist, he had been hired with the reputation of being the leading American exponent of the "New Botany," the study of plants by microscope and experimentation in the laboratory. At Ames, he had set up the nation's first microscope laboratory for botany. Bessey was self-consciously an innovator, a radical in science. He had a penchant for "stirring up of the animals," taking a good-natured pleasure in baiting the scientific conservatives.[5] In

1893, while the political radicalism of populism flamed across the midwestern grasslands, Bessey was, as a middle-aged man, still burning scientific radicalism into his students. In the record book of the botanical club of the University of Nebraska, Bessey was entered as imploring of a new student member:

> Dr. Bessey Soc. addressed words of administration to Saunders Nov., reminding him of the "danger of falling into conservatism," of the "desirability of having a natural bias in favor of anything radical," and of the need "of having his brain and his mind, theoretically and practically, all his life in a meristem state, and of guarding against the growth of much permanent tissue."[6]

His radicalism was rooted in refusal to continue the systematic studies and classification of plants which had been central to traditional botany. He believed in the microscope, in physiological experimentation on the plant, in laboratory-tested facts. He did not hold on dogmatically to theories and willingly took up new and controversial theories, even if tradition cautioned against them. He had no sense of botany as an abstract or museum science. His entire career was taken within the sheltering justification of the practical benefits of science. Though he did not think that any science would prosper if forced to specify in advance what practical benefits would flow from research, he did not doubt that practical benefits would result. He once attributed his radicalness to a life in the provinces of the Midwest and in the very provinciality of its new colleges and universities. The "New Botany" may have originated in Europe, but C. E. Bessey shared none of the hauteur of the German professor, little of his distance from the needs of the people.[7]

Botany for High Schools and Colleges exemplified Bessey's tradition-breaking approach to botany and established his reputation.[8] The textbook was based on the European primer of the new botany, Julius Sach's *Lehrbuch der Botanik*. Bessey requested his publisher, Henry Holt, to mail copies of *Botany for High Schools and Colleges* to botanical colleagues and teachers around the country. Praise was not wanting. The book was "full of the spirit of modern science."[9] "Nothing in the English language" could take its place as a presentation of structural botany based on the compound microscope.[10] It was "nearer the ideal text-book of botany for our American colleges than any other one that has yet been offered."[11] R. N. Prentiss,

Bessey's teacher of botany at Michigan Agricultural College, who had by 1880 moved to Cornell University, must have pleased his former student with a firm "excellent."[12] Asa Gray had not had the opportunity to read it carefully, yet felt certain enough of it to pronounce the restrained approval, "very credible."[13] The historian who has written more on American botany than any one else, Andrew Denny Rodgers III, has reflected the praise of Bessey's contemporaries for the *Botany* in unequivocal words: "Without exaggeration, it may be said that Bessey's *Botany for High Schools and Colleges* was one of the most important works ever published in the annals of the botany of the western hemisphere."[14]

Immediately, the young botanist entered into the top rank of plant scientists. He was offered—and accepted—the editorship of the botanical department of the popular magazine, *The American Naturalist*.[15] This national magazine was guided by Edwin Drinker Cope (1840–97), a famous and controversial paleontologist on the faculty of the University of Pennsylvania. Continuing visibility and influence within the profession were assured. Bessey resigned this subeditorship to move, in 1897, to the botanical editorship of the prestigious *Science*. In the 1910 edition of James McKeen Cattell's *American Men of Science*, he was listed as "starred," one of the one hundred most important contributors to American botany, an honor conferred by ballot by botanists.[16] Despite this Bessey never became the nation's greatest botanist, as might have been signified by a move to Harvard as successor to Asa Gray or to Columbia University— both traditional centers of botany. He poised on the verge of this honor, frequently accepting election to offices of professional societies: presidencies of the Microscopical Society (1901), the Society for the Promotion of Agricultural Science (1889–91), the Wild Flower Preservation Society (1903, 1908), and the Botanical Society of America (1896); secretary and vice-president of the Forestry Association of America (1884, 1893, 1902, and 1907).[17]

Bessey also moved his influence outside the worlds of the university and of professional science, especially in high school curriculum reform and nature conservation. Although not particularly related to each other, these interests were indirectly tied by his professional career. As an administrator at Iowa and as dean of the Industrial College at Nebraska, he was concerned with the high school science preparation of his entering students. In Iowa, he was a member of the advisory committee of the Superintendent of Public Instruction.[18] In 1892 he was a member of the conference on natural history, which had been set up as part of Harvard President Eliot's effort to reform and standardize high school instruction.[19] In 1897 he represented botany on the Committee of Five in the National Education Association, which made recommendations regarding the organization and intellectual standards of secondary school science for college preparation.[20] Toward the end of his career, he was a member of the "Committee on the Definition of the Unit in Botany" of the Association of Colleges and Secondary Schools of the North Central States, which concerned itself with the narrow problem of accreditation of secondary school botany for entrance into college.[21]

Memberships on educational committees were to be expected of a well-known scientist, but Bessey's concern for high school science was better exemplified in his many thoughtful, long letters of advice to high school teachers of botany. While I have made no count of these numerous letters, the care he gave to them in the midst of a correspondence of perhaps 30,000 letters demonstrated a significant commitment at the center of his life.[22]

Bessey also held a lifelong interest in nature conservation. He was responsible for initiating forest conservation practices in Nebraska, and he was active in national preservation organizations, as his several elected offices indicated. Bessey involved himself in the political movements for forest conservation in the sand hills region of Nebraska, Calaveras County, California, and the Appalachian Mountains of the East. In 1887 he eagerly accepted appointment to a joint committee of the Nebraska State Horticultural Society and the State Board of Agriculture, formed to encourage the legislature to reforest the sand hills. Although this effort failed, Bessey successfully interested Dr. B. E. Fernow, director of the Division of Forestry of the Department of Agriculture (USDA), in an experimental reforestation in the region. After a donation of land and seedlings in the early 1890s, the USDA agreed to supervise the project. It was not until 1902, however, that large-scale reforestation was begun. Residents of the area petitioned the USDA in 1900. Bessey solicited the support of Gifford Pinchot. Finally, President Roosevelt in 1902 created two forest reserves in the sand hills region.[23]

After election to the board of the Wild Flower Preservation Society in 1902, Bessey entered the controversy over preservation of the giant sequoias in Calaveras and Tuolumne counties, California, which were threatened by logging. The sequoias were not saved until 1931, when the state of California purchased the Calaveras groves.[24] His last major effort for conservation came in 1907, when he joined the fight in the East for an Appalachian national forest. He successfully obtained resolutions from the Nebraska Park and Forestry Association and the State Horticultural Society in 1908 favoring the forest reserves, and for several years lobbied Nebraska's state congressmen to support the reserves. His national reputation brought him appointment as chairman of the Committee on Education of the National Conservation Congress in 1912.[25]

Bessey's activities in conservation, the high school movement, the transformation of the new University of Nebraska and the New Botany provided his botanical work with a set of linked contexts. Botany was not an "ivory-tower" discipline; he did not expect botanists to be passive. Scientific research led outward to public service, even to utilitarian problem solving. The scientist who followed Bessey's example moved continually through the realms of research, administration, reform, and politics. Scientific knowledge was not isolated.

"*Canis* Pie"

Charles Bessey's greatest accomplishment was more than the sum of his separate accomplishments in the New Botany, educational reform, and conservation. It was the creation of a spirited, educationally progressive, professionally advanced botanical "school." The students ranged from undergraduates in first-year botany to postgraduate scientists. In its years of intense development from 1886 to 1907, this school compares with the major peacetime scientific laboratories and research groups in our nation's brief history of science—Louis Agassiz's laboratory of zoology in the 1860s at Harvard, Thomas Hunt Morgan's chromosome research group at Columbia in the 1910s, E. O. Lawrence's cyclotron laboratory at the University of California, Berkeley, in the 1930s, for instance. The students and scientists who participated in the school and grew under Bessey's inspired tutelage, such as Roscoe Pound, Frederic Clements, H. J. Webber, among others, warmly remembered the years as decisive to their scientific careers. Their friendships and collegial associations persisted through two generations into the 1950s. The Botanical Seminar, as they were informally known by 1886, was a true "invisible college," flourishing outside official academic status as a club, independent of the Department of Botany, with the dedicated ardor of a social fraternity. For a generation of Nebraska students, the "Sem. Bot." mediated between the pupilage of the classroom and the professionalism of research botany. Seminarians met regularly for the reading of scholarly papers and discussion, sponsored botanizing excursions in the neighborhood of Lincoln and longer trips during the summers, and published research reports (eventually with appropriations from the state legislature). They established and ran the Botanical Survey of Nebraska in the 1890s, providing valuable professional experience for nascent biologists and extensive information that would enter the first formulations of grasslands ecology. In imitation of the social fraternities of the day, Sem. Bot. had initiation ceremonies and semisecret ritual, replete with fraudulent Latin, dreamed up by Roscoe Pound. Ranks differentiated the membership, and serious examinations in botany were required for passage from novitiate to the higher grades. It was a scientific brotherhood, a seminary for the missionaries of botany.[26]

Since the Sem. Bot. was the key institution in the Bessey milieu within which grassland ecology originated, we can obtain an understanding of the emergence of this specialty by examining its origin and early history. Bessey's arrival in Lincoln, Nebraska, in 1884 coincided with the matriculation in the university of several students who later became prominent botanists. The most enterprising of these was Roscoe Pound. Pound received his baccalaureate in 1888, and went for his master's degree, which he took in 1889. While an undergraduate, he spent much time in Bessey's botanical laboratory, to which he returned in 1888–89 as assistant in the botanical lab, while working on his advanced degree (on the subject of the imperfect fungi of Nebraska).[27] From at least fall term, 1886, Pound's closest friend in the botany lab was H. J. Webber, son of a farmer outside Lincoln. Webber, who entered the college section of the university in 1885, found himself in the botanical laboratory with Pound in 1886. They became high-spirited friends. Webber later worked for the USDA and the University of California, eventually becoming director of the university's Citrus Experiment Station at

Riverside, in the 1920s. He has left us the longest personal memoir that I have been able to locate of the Bessey botanical laboratory of the 1880s; though anecdotal, Webber's memoir reveals clearly the spirit that made the laboratory so formative.[28]

Not only was the University of Nebraska a new school in the 1880s, but Lincoln, the state capital, was similarly new. The city and the university were small, and photographs of the decade reveal a settlement of about 10,000 persons (1880) domiciled in scattered homes that wandered without boundaries onto the prairies, dirt streets unshaded by large trees. In 1884, the university consisted of but one naked building. The school enrolled 373 students, with perhaps half of them in the preparatory Latin School rather than the college proper.[29] By the end of the decade, four new buildings were added, enrollment increased by a third, and a new graduate program produced a few master's degrees. Bessey's botanical laboratory was completed in April 1885 and moved to the new chemical building, apparently in early 1886.[30] For the students, life at the university centered around their classrooms; for botany students, Bessey's new laboratory took on the character of a social center. Not only did the students spend much time in the laboratory conducting research but gathered there also for conversation and pranks. Literally, there was nowhere else to go.

Even in the early 1890s, when the university was more than two decades old, there were no dormitories or resident fraternities. Unless their families lived in Lincoln, students boarded around town—"where we could and as we could," Willa Cather wrote.[31] Without the random social mixing provided by official residence halls, students tended to socialize formally along lines of academic studies to a greater extent than did later students. Student organizations were divided between literary societies and scientific study groups. The literary groups (Webber remembers two during the 1880s, the Union and the Paladian) met weekly for debates, papers, and socializing. Webber confessed to obtaining most of his social skills at these meetings. Among science students, informal cliques seemed to dominate, until the emergence of the Botanical Seminar, which was formally organized in 1888. Bessey's botany students were clearly among the most coherent of these science-student cliques. Webber and Pound both recalled later a healthy rivalry, usually involving pranks, between the literary students, whom they

called "lits," and the science students, whom they called "sci's." Of
the years between 1885 and 1888, Webber wrote:

> Led by a clique of the botanical students who dominated the spirit of
> the group, the scientific students, or "sci's," were gradually molded
> into a sort of fighting organization to ballyhoo (?) [*sic*] for science. We
> had no organization, but all responded faithfully to certain unwritten
> ideas, mainly to protect every scientific student from the indignities
> that might emanate from the classical and literary students, all of
> whom we dubbed "lits." We developed the habit of exhibiting our
> disdain for such students by catching them, one or two at a time, and
> tossing them into the air.[32]

Webber remembers no effective return pranks by the "lits"—"It was
a surprise to us that the literary students did not organize the students of all the nonscientific departments to retaliate on us, but they
evidently lacked leadership."[33] Probably this was convenient forget-
fulness; certainly the lits were capable of pulling one on the scientists. A few years later (after Webber had left), Willa Cather
sponsored a retaliatory trick. Cather attended the university from
1890 to 1895 (requiring two years in the preparatory school), and was
a close friend of Roscoe Pound's two sisters, Louise and Olivia. She
and another friend learned in the fall of 1892 that the members of the
Botanical Seminar intended to surprise Bessey with a gift, a bust of
Charles Darwin. The two young women hid the bust, leaving the
science students to believe it was stolen. After the scheduled hour of
presentation to Bessey had passed, Cather's friend "found" the portrait sculpture of Darwin, and allowed the Bessey students to think a
man had stolen it.[34]

Bessey's allocation of space in the old chemistry building was the
natural social center for his botany students. His office, the botany
laboratory (with microscopes and reagents), and the herbarium were
physically contiguous in the building and it was possible to be in one
room and hear Bessey walk in a nearby room. Webber, referring to
himself, Pound, and a few other advanced botany students, called
the herbarium "our ordinary meeting place."[35] On one occasion,
following a prank that involved public labeling of university faculty
with spurious and sarcastic Latin names invented by Pound, the
culprits retreated to the herbarium and listened to Bessey's footsteps
advance down the laboratory, his manner brisk with admonition.[36]

Webber offered another anecdote to the spirit of the laboratory. One afternoon, he began flipping popcorn kernels at students. The resultant good-natured foolery led Webber to wonder why popcorn pops, a question he asked Bessey in the next day's lecture. Bessey, no doubt bemused, threw the question back, challenging Webber to investigate it. Webber subsequently spent hours in the laboratory, examining popcorn sections under the microscope, testing tissue with stains, and concocting an explanatory hypothesis. Bessey may not have recommended publication, but Webber remembered the incident vividly for what it told of the pleasant freedom and continual research challenge of the Bessey spirit.[37]

According to reminiscences later, the Bessey students were calling themselves the "Botanical Seminar" as early as 1886, but, in fact, the group did not formally organize until the fall of 1888. While the high color of the Webber memoir does not allow us complete certainty, apparently much of the pseudoritualistic content of the formally organized Botany Seminar in 1888 was a carry-over from the previous two years. Even the seminar's "song," "*Canis* Pie" (or dog pie or mincemeat pie), antedated formal establishment of the Sem. Bot. The question is why the step to formal establishment should have been taken. I have been unable to locate any direct testimony from the Bessey students, but when we examine what was happening to the individual students and to Bessey, a persuasive explanation emerges. Until the fall 1888, the group of Bessey students was held together by their formal courses of study, including laboratory study under Bessey. In the new academic year, 1888–89, the teacher and his students took different directions, which destroyed the informal clique. Formal organization served to bring the group back together regularly. According to the Record Book of the Botanical Seminar, in the academic year 1886–87, five students were together in Bessey's lab: J. G. Smith and J. R. Schofield, both juniors, and H. J. Webber, T. A. Williams, and Roscoe Pound, all sophomores. The record book noted, "Here there sprang up between them a lasting friendship out of which grew the Sem. Bot."[38] We have seen from Webber's memoir that he and Pound were fast friends, and the record book adds that the friendship was more than a partnership for highjinks: "Mr. Webber and Mr. Pound began to collect together extensively."[39] In the following academic year, 1887–88, three new sophomores were added to the group, A. F. Woods, T. H. Marshland, and L. H. Stoughton. The group's song, "*Canis* Pie," was written, probably by H. J. Webber, who invented the practice of eating pie at meetings.[40]

In the fall of 1888, this enterprising group of young scientists nearly disintegrated. Their presiding spirit, Bessey, was appointed acting chancellor while the university selected a new chief administrator. He was frequently absent from the group and no doubt this was traumatic; the record book of the group reflects the importance of this absence by noting four years later that in 1888, Bessey "was necessarily absent a large portion of the time."[41] Some of the students also changed their schedules. Smith was graduated in the spring, was appointed assistant to the Agriculturalist, and was largely at the Experiment Station doing graduate work. Schofield was briefly absent (he would return by November) at Geneva, Switzerland. Pound was appointed assistant in the laboratory, but was beginning his master's degree research on the imperfect fungi of Nebraska, which no doubt pulled him away from the campus for short periods of field collecting. Webber was a senior and was preparing a catalog on Nebraska flora which would be the basis of his master's thesis, written the following year. Some of his study was undertaken in the herbarium, but he must also have scoured the environs of Lincoln and eastern Nebraska, botanizing at train stops, which also would have taken him away from the laboratory. Webber was absent for the entire summer of 1889, traveling throughout Nebraska with Dr. Lawrence Bruner, a university entomologist. In this context, formal organization was a natural means of keeping the group together. They gathered weekly on Thursday evenings in the herbarium for papers and criticism. Presumably, Bessey was able to attend the evening meetings.[42]

Formal organization also served a professional purpose. Since by 1888, several of the group already were in graduate study and Webber was to begin the following year, the group, for the first time, would have wanted a professional organization to rehearse the fundamental professional activity of presenting contributions to knowledge to a group of scientific peers for critical appraisal. They were leaving the status of students and entering the realm of the independent professional. They were nearly the very first to do so at Nebraska. Isolated by two thousand miles from eastern professional societies and the traditional centers of academic botany, they created the scientific society they needed in Lincoln. The seriousness with which they took these meetings is indicated by the format: topics of papers were announced six weeks in advance and seminarians were expected to read in preparation. Between October 11, 1888, and April 18, 1889, twenty-one papers were read, on topics ranging from "The Extent and Position of the Group of Slime Moulds" (Pound) to "The Doctrine of Pangenesis [Darwin's genetic theory]" (Smith). With such a rigorous schedule, no wonder they felt they deserved a picnic on April 25, and the "traditional" mincemeat pie.[43] The central role of Pound in this group is indicated simply by the comparison of the high intellectual activity of the academic year 1888–89 and the less intense activity of the following year. Roscoe Pound spent the academic year 1889–90 studying law at Harvard Law School, and no formal papers were read by his botanical compatriots remaining in Lincoln. As the record book reads: "The work [took] the form of informal discussions and examinations of current literature."[44]

Members of the Sem. Bot. further formalized their organization in 1892, by accepting new undergraduates, who were given the status of *novitii*. The first such admittee was the sophomore, Frederic Clements. The original 1886 group of Bessey students was comprised of undergraduates, of course, but since June 1889, all members were of graduate status. The motivation for this step is not difficult to discern. Key members of the group had moved away from Lincoln after 1890. In the summer of 1890, Webber studied at the Woods Hole Oceanographic Laboratory (which was then in its third year of operation) and returned to Lincoln only briefly to be married. He began study for his doctorate at Washington University in Saint Louis that October.[45]

Pound was no longer a degree-pursuing graduate student. He passed the Nebraska bar examination in 1890, after returning from Harvard, and was hired as a stenographer by a Lincoln attorney. In 1892 he became a partner and, presumably, his responsibility and the demands on his time increased.[46] His collecting expeditions were generally limited to weekends and vacations (little collecting was done in the winter in Nebraska, of course) and botanical discussion was carried on in the weekly Sem. Bot. meetings. T. A. Williams had left Nebraska to be an instructor at South Dakota Agricultural College. L. H. Stoughton had gone to Harvard Divinity School. T. H. Marshland remained in Lincoln, but was teaching high school, rather than a graduate student. J. G. Smith had gone on a world tour. In the fall of 1891, therefore, only four members (Pound, Marshland, Woods, and Schofield) of the group of fall, 1890, remained active. Clearly the group had to admit new members to remain vital.[47] Consequently, Per Axel Rydberg, who was a Bessey graduate student, was admitted in the fall of 1891, and Clements was admitted in the spring of 1892, at which time the group was opened generally to undergraduates who could pass an examination. The Botanical Seminar was now provided with a mechanism for perpetuating its membership. It had the quality of a permanent institution. To signify this new status, Pound and Smith compiled the history of the previous four years of the seminar's existence, entered as the first twenty pages of the record book: "Pre *Canis* Pie," the era was labeled, no doubt by Pound.

The Bessey System

The history of the Botanical Seminar reveals clearly the peculiar qualities of "the Bessey system," as the instructional and research context established by Bessey was locally named. The intellectual paradigm for ecology that emerged at the University of Nebraska in the 1890s among the scientists of the Botanical Seminar was intimately involved with this vibrant milieu of laboratory research, the "new" scientific botany, concern with practical problems of agriculture, interdisciplinary contribution and criticism, and the multiple perspectives of scientists from different vocations—law, teaching, administration, and federal research in agriculture. To isolate any one component of this setting, such as student training in the laboratory method, as several historians have done, is to miss the texture and feel of the daily scientific life of the group that promoted perception of scientific problems that other scientists had not seen and innovative solutions to those problems. The lives of these scientists were not neatly compartmentalized.[48]

The style with which Bessey successfully pursued his different roles was an enduring inspiration to his students. They might not be able to match him in this regard, but they would be led by their admiration of him to accept the legitimacy of the open scientific setting and of radicalism in science. This radicalism in science did not necessarily mean they were rebellious in politics or social activities, of course. Webber remembered back over sixty years, being a radical in politics, participating in the populist movement in the 1890s, but Pound was a most conventional lawyer in the 1890s, and had only scorn for the Populists.[49] Bessey was particularly attuned to the botanical problems of Nebraska's farmers, and his botany was never pure of these practical concerns; science did not mean shutting out problems of everyday life. As the official, albeit unpaid, state botanist, Bessey repeatedly encountered problems that today would be described in terms of the ecology of invasions. He tried to advise farmers as much as possible through correspondence but occasionally he took to the fields to see for himself their problems, as during the invasion of the Russian thistle, a tough, strongly rooting tumbleweed that took over croplands in the late 1880s. The correspondence on economic botany became so heavy that in 1898, Bessey complained to the regents of the university. "In spite of the fact that no effort has been made to attract correspondence upon problems connected with the economic aspects of botany, I find every year that the answering of questions by correspondence is a heavier and heavier burden."[50] His students were frequently directed toward economic problems of botany or toward botanical questions relevant to agriculture. Webber and Per Axel Rydberg were both sponsored in surveys of the state's flora by the Department of Agriculture.[51] After the Bot. Sem. undertook the systematic botanical survey of the state in the 1890s, legislative appropriations aided the enterprise at one point.[52]

Bessey's correspondence indicated the ready manner in which he made botany a resource of the agricultural experiment station.[53] As a resource, however, botany had to be expanded into a set of new problems that required additional books and journals for the department library, different collections for the herbarium, and new experimental apparatus. Ultimately, the new problems required a new intellectual framework for botany, and at this point, the science of ecology emerged. We can see how the new problems were pushing Bessey to the threshold of a new approach to the vegetation of the state. In a report as Botany Department chairman to the regents of the university, Bessey mentioned that the introduction of irrigation in the state had brought a host of new problems: weeds not previously found and fungi not before encountered in the dry summer plains both appeared. He had been forced, he told them, to anticipate that these problems would arise and build up the botanical department's resources to deal with them in the future.

When the time comes for taking up the lines of botanical investigation demanded by the general introduction of irrigation, when the harmful fungi have so increased as to require particular attention, and when a demand is made for an investigation of the habits and life history of the weeds of the state, or the usefulness and means of propagation of our native grasses, the department of Botany will be found ready.[54]

2
THE CREATIVE TENSION:
Laboratory Versus Nature Study

The Return to Nature

The introduction of the New Botany into colleges and high schools in the final decades of the nineteenth century coincided with a national outburst of enthusiasm for the out-of-doors and nature study. The call for preservation modulated the historic American will to subdue the wilderness, to transform it into the garden or industrial landscape. With little desire to exploit the land as farmers or foresters, the new middle class nonetheless yearned for a redeeming contact with nature. The symptoms of the new national mood were varied: streetcar suburbs and urban parks for refuge from the dense, angular, chaotic city; Fresh Air Camps for children; the Ornithological Union (established in 1883) and the National Association of Audubon Societies (1886); admired nature interpreters, such as John Burroughs and John Muir; the Adirondack wilderness of one million acres, constitutionally preserved in 1894; Arbor Day; preservation societies (Appalachian Mountain Club, 1876, and Sierra Club, 1892). Concern for disappearance of the wild melded with concern for the disappearance of rural life in the rapidly urbanizing nation. Theodore Roosevelt's muscular advocacy of the strenuous life made him the appropriate president to appoint the Country Life Commission, whose objective was to study the rural farm community, if not the "Boone and Crockett" individualism he espoused.[1]

Charles Bessey hoped the fad could be channeled as support for scientific botany in schools, which in turn might produce teaching jobs for his young botanists. As in most academic efforts to capitalize on cultural fashion, risks were involved. Nature appreciation was a diffuse affectation; scientific study of plants and animals required discipline. Nature study promised a quick route to nature knowledge; scientific knowledge required years of study in a structured program. Nature study was ideologically antitechnological; the New Botany was based on sophisticated and expensive microscopes and on chemistry. Nature study could easily supplant scientific study in high schools. In academia as in economics, bad coin drives out good.

As it was to develop in the 1890s, the science of ecology was closer to nature study than the other natural sciences. Frederic Clements briefly tried to ally his new science with the fad. He adapted botany to field study for high schools. John Merle Coulter also urged in the late 1890s that the general study of plant life, as in nature study, introduce students to botany. To these flirtations Bessey reacted negatively. He considered it yielding to dilettantism and an abandonment of the search for knowledge. It was a false foundation for the study of plants. Bessey's reaction illustrated the dilemma in which the scientists found themselves, torn between the shimmering opportunity to promote their own interests and the fear of debasing the standards of their work. The tension was nevertheless productive. It motivated a rigorous foundation and mathematicization of grasslands ecology.[2]

No single event or set of events appears to have catalyzed the explosion of general interest in nature; yet, the interest is demonstrable. One index of eruption was the rate of establishment of new magazines and journals devoted to natural history or to special topics like birds and plants. Figure 1 shows the numbers of new, small circulation, short-lived, natural history periodicals. These little journals more nearly represented grass roots enthusiasm than any other form of publication. In the years between 1884 and 1891, they blossomed like flowers after a spring rain. And no doubt the depression of the 1890s was responsible for drying up support for them.[3]

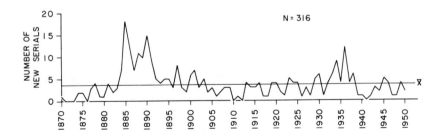

SOURCE: Margaret H. Underwood, *Bibliography of North American Minor Natural History Serials in the University of Michigan Libraries* (Ann Arbor: University of Michigan Press, 1954).

Fig. 1. New minor natural history serials

For the surface of this phenomenon, at least, an explanation is not difficult. The ideological content of the back-to-nature movement was filled with antiurban sentiment. Presumably discontented urbanites filled the ranks of bird watchers and backpackers, and sent their children to the summer camps. Peter Schmitt, in *Back to Nature*, explains that "an increasing number of city dwellers turned 'back to nature,'. . . mainly to escape the minor irritants of urban life."[4] The city was the source of familiar corruption. Children raised in cities became worldly and sophisticated at too early an age, were oriented toward facile cleverness, while at the same time they lacked self-reliance. Adults were continually stimulated to reach unobtainable goals and wealth and never found contentment. The dense cityscape closed the distant vistas that drew men toward ideals of grandeur and largeness of moral vision. Social relationships were based on second-ary associations, such as membership in clubs and work roles; the primary face-to-face relationship was impossible when human inter-action was limited to brief encounters on streetcars. Noise, imper-sonalization, impoliteness: all created strains that often led to neuroses. American historians are no doubt provincial not to see that these complaints are not original. In pure literary tradition, the antiurban literature is traceable to Roman satirists like Juvenal, who was repulsed by the moral and cultural decay he perceived in Rome. More immediately, a vigorous literature of antiurban ideology had flourished in England since the late eighteenth century, providing much of the thematic content utilized by Americans two generations later. Indeed, many of the British organizational responses to the industrial city were exported to America, such as the Boy Scouts, the Christian youth organizations, nature study groups, urban parks, and the garden city movement.[5]

The antiurban literature served a greater purpose than to salve sensitivities bruised by the scramble of city living. In Great Britain, as Raymond Williams has recently argued, antiurban writing was sponsored and patronized by a new class of rural entrepreneurs who were eager to cover their exploitation of the rural poor by pointing to the exploitation and moral degeneracy of the urban wealthy.[6] We may similarly discover ideological purpose behind some of the nature study activity and publications in the towns and cities of the United States. Towns and small cities were dynamic centers of American culture in the last two decades of the nineteenth century. Towns produced more botanists per capita than either true rural areas or true cities. While eastern cities expanded rapidly, many towns came into existence as population booms cyclically swept across the Midwest and the Pacific Coast. These new settlements were creating cultural institutions for the first time—libraries, fraternal organizations, literary societies, schools, entertainment halls. Nature study may well have been, verbally, antiurban; it was also part of the vigorous formation of culture. It was middle-class cultural activity, typical of the middle class portrayed for us by sociological novelists like Sinclair Lewis and sociologists like Herbert Gans, typical both of towns in the late nineteenth century and suburbs after the Second World War: organizing, meeting, promoting, moralistic, printing little newsletters, depending on the enthusiasm of members rather than on direct economic interest. Back-to-nature not only meant escape from the minor irritants of urban life; it affirmed urban life by building urban life.[7]

The Promise of the High School

The dramatic rise of the land-grant college and of graduate education in the last quarter of the nineteenth century has generally hidden the intellectual importance of the equally dramatic rise of the high school. The high school was the educational institution where the nature-study movement and the hopes of Bessey and other natural scientists met. In the 1880s, Bessey introduced laboratory instruction into high schools and in the following decade he participated in the important effort, led by scientists, to modernize the secondary curriculum.[8] Several trends encouraged Bessey and other scientists to hope that a large future existed for the disciplinary sciences. First, public high schools came clearly to dominate secondary education, and they were subjected to state regulation that raised their minimum standards and coaxed uniformity out of their chaotic diversity of the early 1880s. Private academies of course were outside public regulation. Contemporaries referred to the phenomenon as the "high school movement." The absolute number of high school students studying science (in preparation for college instruction) increased by a factor of five between 1882 and 1900 (see table 1). Nebraska, a boom state in the decades of the 1870s and 1880s reflected the national

Table 1
Enrollment Statistics for High School Science Classes

School year	Number of high schools reporting	Total enrollment	Number of students in sci. prep.	Relative % in sci. prep.
1871–72	811	98,929	992	1
1882–83	1,482	138,384	3,566	3
1899–1900	6,005	519,251	24,919	5

Source: *Report of the Commissioner of Education for the Year 1872* (U.S. Office of Education, Annual Report, U.S. Government Printing Office), pp. xxv–xxxvi; ibid., 1882–83, p. cx, Table I; ibid., 1899–1900, vol. 2, p. 2122.

trends (see table 2). Second, the Latin curriculum of the high school was gradually supplanted by the English curriculum, though instruction in the ancient languages remained heavily enrolled past the turn of the century. In this curriculum shift, instruction in the sciences was expanded drastically, under the general rubric of "preparation for life." Indeed, scientists were largely responsible for the modernization of the high school curriculum, which saw not only the introduction of science education but also history, English language and modern literature, and modern foreign languages.[9]

A detailed study of the nineteenth-century curricula of north central states' high schools, done by John Stout after the First World War, reveals precisely the changes in the high school instructional programs. Drawing on school board reports and instructional materials, Stout was able to reconstruct the curricula of more than forty high schools from 1860 to 1918 (though he was not always able to reconstruct the curriculum for each high school over the whole period). He shows "on the whole, the range of work provided was comparatively narrow until about 1885, except in a few of the larger

Table 2
Nebraska High Schools

School year	Number of high schools	Total enrollment
1870	3	161
1880	40	*
1918	442	29,273

*Not available.

Sources: U.S. Bureau of the Census, U.S. Census for the Year 1870, Vol. I, *Statistics of the Population of the United States . . .* to which are added the statistics for School Attendance (Washington, 1872), Table XIII, p. 466, Table V, p. 916; U.S. Bureau of Education, Biennial Survey of Education, 1916–18, Bulletin (1919) no. 90 (V. III), table 14, p. 37, table 40, p. 166.

schools."[10] English frequently was not offered at all before 1885; science instruction tended to be in the natural history subjects—geology, geography, "natural philosophy"—with a surprisingly high percentage of the schools (over 65 percent) offering chemistry. Few schools offered any history at any time. Not until after 1885 did more than half of the high schools surveyed offer United States history.[11] The core of the high school instructional program until 1885 was in Latin and arithmetic and algebra.[12]

The brief decade and a half between 1885 and 1900 were the golden years for the unalloyed modern curriculum, centered on English, the modern foreign languages, history, and the disciplinary sciences. English grammar and English (as contrasted to American) literature were offered by more than half the high schools. The disciplinary sciences, particularly botany and physics, supplanted natural history and natural philosophy, which virtually disappeared from the schools by 1900. Greek was largely removed. Latin, interestingly, was increasingly emphasized; in the 1860 to 1865 period, only 5 percent of the schools offered four years of Latin, while in the 1896–1900 period, 75 percent offered advanced Latin.[13] In Lincoln, Nebraska, high school, for instance, botany was taught for only one-third of a year in the 1881–85 period, while a decade later, 1895–1900, it was being taught for one-half of a year.[14] Bessey and his students were involved in this shift, as we shall see below.

The instructional content of science courses demonstrated the shift to disciplinary science, away from natural history. In botany, Asa Gray's *How Plants Grow* (first edition, 1858) was the most widely used textbook; and even in the 1890s, half the schools continued to use the book. At the same time, however, newer textbooks were added, reflecting the rise of the New Botany of the 1880s. J. Y. Bergen's *Elements of Botany* was widely adopted after 1896. Gray's text explicitly reflected the deism of its author, quoting Bible passages, and referring to the design of nature by God. The scientific content of the textbook was restricted mainly to anatomy of plants and classification. Bergen's textbook, in contrast, emphasized physiology and morphology of plants, utilized microscopic examination of plant structure, and discussed the plant in its environment in terms of Darwin's theory of struggle for survival and natural selection. Bergen explicitly developed the classification of plants from the point of view of Darwin's theory of evolution. He referred to the botanical textbook of Eugene Warming, one of the founders of European ecology, for a full analysis of the relationship between plants in the classification of species. Bergen's was an excellent scientific textbook for the time and would be credible a half century later.[15]

These educational shifts in the high school curricula were reinforced by the Report of the Committee of Ten of the National Education Association in 1893. The committee had been called in 1892, to recommend uniform standards for high school curricula. The call grew out of continuing consternation among colleges regarding the lack of quality and uniformity in college preparatory instruction in high school. The committee was named at the urging of Charles William Eliot, president of Harvard, at the 1892 meeting of the National Education Association in Saratoga, New York. Eliot chaired the committee and wrote its report, thereby making himself the most influential member of the committee.[16]

The strategy of the Committee of Ten was to appoint advisory committees in each area of the high school curriculum, with representatives from around the nation. The advisory "conferences" were instructed to answer a list of questions, such as " 'at what age should the study which is the subject of the Conference be first introduced?' " and " 'What topics, or parts, of the subject may reasonably be covered during the whole course?' "[17] The "Natural History" conference, which was to review biology, botany, zoology, and physiology, included C. E. Bessey and John Merle Coulter (then president of Indiana University in Bloomington); the New Botany was strongly represented.[18]

The Conference on Natural History recommended the introduction of botany and zoology in the primary schools, with instruction for "not less than two periods a week, throughout the whole course below the high school."[19] Laboratory work on plants and animals in high schools was described as an "absolute necessity."[20] Laboratory notebooks should be kept. Examinations should include written tests and laboratory tests. It is interesting that the final report of the committee did not mention any necessity for fieldwork for botanists. Nevertheless, the botanists, Bessey and Coulter, certainly believed field studies were a part of the secondary course in botany. Privately responding to inquiries on his own instruction in botany, Bessey mentioned that he sent his students into the field as soon as weather permitted.[21] Field study was also mentioned in a published syllabus for botany by Bessey in the *Northwestern Journal of Education* (1891–92).[22] The absence of field study from the final report, therefore, did not reflect heavy antagonism toward it by the botanists. Rather, it reflected the ideological thrust of the report, which was toward *disciplined* scientific study. Disciplined scientific study occurred in the laboratory, whether in physics, chemistry, or botany. It also reflected the power of the New Botany, then at its peak, just before the "nature-study" movement was to break out.

After 1900, the position of the disciplinary sciences in the high schools seriously eroded, so seriously that a mere two decades after the report of the Committee of Ten, Bessey lamented that perhaps he and the other founders of the New Botany had pushed laboratory studies too strongly. A variety of changes in secondary education worked against disciplinary science. The laboratory was expensive, and many school districts could not afford the necessary microscopes and other equipment.[23] An analogous problem was securing trained teachers in the disciplinary subject matters. In 1905 a survey revealed that 70 percent of the male teachers and 53 percent of the female teachers had graduated from college, while in some states as few as half of the high school teachers had graduated from college.[24] It was understandable therefore that schools quickly turned to general science courses and nature study courses in which science material could apparently be taught with less qualified staff. The first "general science" course appears to have been taught in Lincoln, Nebraska, and the course quickly spread nationwide after 1905.[25] While the general science course did not necessarily supplant disciplinary science courses, it undoubtedly satisfied the science requirement for larger numbers of students.

The strongest force undermining the disciplinary sciences was the rise of business and vocational training in the Progressive era. The genteel tradition's notion, as expressed by Eliot, that disciplinary science and historical and literary scholarship were the best mental training for life, was challenged and simply rejected by much of the public. The vocational movement spread like a "mental epidemic," after the Massachusetts Commission on Industrial and Technical Education called in 1906 for training the "industrial intelligence."[26] The movement was given official recognition the following year, when President Theodore Roosevelt picked up the call for industrial and agricultural education in his annual message to Congress.[27]

The relative shift of high school resources and the cultural shift of American public attitudes after 1900 took their toll on the high school sciences. Enrollments in science courses plummeted, in comparative terms. In 1910, the sciences enrolled 82.7 percent of high school students; in 1915, they enrolled only 65.4 percent. The enrollments of botany and zoology collapsed; their combined relative percentage of 24.9 percent of the students in 1910 fell to 12.4 percent in 1915.[28] In the colleges, Bessey complained, it was becoming impossible to interest botany students in becoming teachers. The rising expectations of the 1880s and 1890s were now sadly unmet.

Natural History in the Genteel Tradition

By the last quarter of the century, American bourgeois culture had settled around a comfortable cluster of values that has been called "the genteel tradition." Genteel culture was primarily literary; print, of course, was the expressive medium of the nineteenth-century middle class, but the tradition also included artistic and architectural styles. The most important value in this tradition was "idealism," defined to mean the existence of nonmaterial or nonnaturalistic standards of quality, such as beauty, independent of human mind but recognized only intellectually. For George Santayana, the Harvard philosopher who originally and derisively characterized the genteel tradition, the idealism was particularly vapid. It was not a theory of knowledge, it was not a systematically articulated ethics; it was so flimsy that it could not bear a formal label, Platonism or Neoplatonism or Hegelianism. It was, nevertheless, quite real, as historians have demonstrated, and dominated the literary establishment of magazines and clubs, the great private universities, and official patronage of the arts.[29]

Howard Mumford Jones perceives the genteel tradition as "an operative fusion of idealism and the instinct for craftsmanship." Not only had literature and art to express universal standards of beauty and universally moral themes, such as the primacy of the home, but the expression had to be executed with obvious skill. Poetry was carefully metrical, painting was painstakingly detailed in line, the short story had literary "effect." Unobstrusive allegory was respected for its indirect reference of universal themes and values;

Augustus Saint-Gaudens's use of allegorical figures, as in his monumental relief of Colonel Shaw, is an obvious example. The craftsmanship was a demonstration of the professional status of the artists. One objective of the exponents of genteel culture had been to nurture an independent artistic community, able to make its living from the sale of art alone. The names of the central figures of genteel culture are now largely forgotten: Thomas Bailey Aldrich, editor of *The Atlantic Monthly* and poet; George Henry Baker, poet; George William Curtis, editor of *Harper's*; Edwin L. Godkin, founder and editor of the New York *Nation*; Richard Watson Gilder, editor of the *Century*; Edmund Clarence Stedman, critic; Richard Henry Stoddard, poet and novelist; Bayard Taylor, playwright. Of course, some of the leading representatives of genteel culture are well remembered, including Julia Ward Howe, the poet, and Charles William Eliot, president of Harvard.[30]

While science was not explicitly central to the genteel tradition, conceived as a literary culture, natural history at least was an important component of the cluster of genteel values. At her home in Boston and later in Newport, Rhode Island, Julia Ward Howe hosted the leading scientists of the day in salons for specially invited guests. Her speakers were a roll call of the prestigious scientists whose work exemplified the genteel attitude in science: Louis Agassiz, the Swiss-born zoologist, who came to Boston in 1846 and cultivated Brahmin support of science at Harvard until his death in 1873; William B. Rogers, president of MIT; Alexander Agassiz, the son of Louis Agassiz, and zoologist and engineer, who headed the Harvard Laboratory of Comparative Zoology; S. Weir Mitchell, a famous Civil War physician, poet, and novelist; Maria Mitchell, director of the observatory at Vassar College.[31] As part of the genteel tradition, natural history was fused with idealism, so that nature was valued for its reflection of the thoughts of the deity who created it. Nature was designed and static in form. Louis Agassiz was one of the major contributors to the "catastrophic" theory of the formation of the earth, for instance; he explained geological history and the succession of paleontological forms in terms of periodic catastrophes that destroyed flora and fauna, with the deity creating life anew in the next epoch. Obviously, he opposed Charles Darwin's theory of evolution, in which chance variation in offspring, naturally selected in a struggle for survival, explained the historic change in forms of species. After the Civil War, other genteel scientists were forced to accept Darwin's theory in some form, but many held onto their deism, as did Asa Gray, the botanist, and Joseph Le Conte, the University of California geologist, by seeing evolution as the gradual emanation of God's creation. Divine creation was seen as a process rather than an act. From the point of view of genteel natural history, therefore, nature could be studied for its manifestation of divinely sanctioned ideals. [32]

By reassuring the genteel culture of scientific support for their vague values, scientists received in return unprecedented philanthropy for their work. The Gilded Age witnessed the construction of the great institutions of natural history, as well as support for the newer disciplinary sciences. The American Museum of Natural History in New York City, the Field Museum in Chicago, the Boston Museum of Natural History, the Missouri Botanical Garden, the Lick Observatory in California, the Yerkes Observatory of the University of Chicago, the Catherine Bruce Photographic Telescope and Endowment Fund for astronomy; the Rockefeller Endowment of the University of Chicago—all were philanthropies guided by entrepreneurial scientists and motivated by their donors' sense that natural history revealed God's plan to men. Historian Howard Miller perceives diverse motives leading philanthropists to endow museums and observatories, not excluding, of course, personal vanity. But they were all aided by their deism, and certainly any suspicion that their endowment would promote materialistic atheism would have inhibited their generosity. [33]

It is difficult to judge to what extent this support of natural history motivated middle class support for specific scientific ideas or inspired students to take up the sciences. It is certain, however, that the new endowments contributed to rising expectations in the scientific community. The observational sciences of natural history were an integral component of the dominant cultural style and values of the nation. Museums, observatories, and laboratories were ranked alongside the new symphonic orchestras, opera guilds, and literary clubs that expressed the Gilded Age's aspirations to high culture. Each major bequest to science was extensively reported and praised by the newspapers and magazines. Bessey and his students in Nebraska, a rural outpost of science, were part of this new excitement over their profession. The leading botanical journal, the *Botani-*

cal Gazette, published by John Coulter, regularly carried news items of support for botany. When Coulter moved to the University of Chicago, the journal moved with him, of course, and therefore reported at length the enormous bequest that established the Hull Botanical Laboratory, the building for which was completed in 1897. *Science*, the news magazine of the American Association for the Advancement of Science, also mentioned state and federal appropriations for scientific study and research in universities and in federal research agencies.

The expansion of the state university systems, the popular "return to nature," the nature study movement, the rise of the secondary school with its successful expansion of the sciences, including botany, and the philanthropic endowment of natural history institutions and research provided a general cultural milieu, in the last quarter of the nineteenth century, for science as part of the cultural establishment. When Bessey told his students to become "missionaries for botany" he knew fully well that the American middle class public was ready for conversion. Bessey and his colleagues in botany were encouraged to promote their disciplines as rapidly as they could. Yet disciplinary professionalism in science presented problems. The research laboratory was not the museum and nature study not the field trip. The enthusiasm of the public for the latter would not magically transform into hard-working discipline at a microscope in a laboratory or sustain years of undergraduate study in a science. By the turn of the century, the public was considerably disillusioned with disciplinary science. And by the First World War the founders of the New Botany came to feel that they had misread the public temper.

The New Botany, the Research Ideal, and Ecology

The establishment of the laboratory in botany was but one part of the widespread adoption of the research laboratory in American science and the rise of graduate education, which the laboratory served. These developments were all intrinsic to the professionalization of science following the Civil War. The laboratory both contributed to and resulted from this sociological change in science. It made possible an explicitness of instruction to the student not typical of the old botany; it defined sequences of problems, progress in which marked the student's increasing mastery of the discipline. The laboratory was not simply a setting with instruments; it implied new sets of problems and a new approach to their solution. That many of these problems were ad hoc practical problems arising out of the agricultural and industrial progress of the age, rather than disciplinary science, only provided the adherents of the laboratory with the opportunity to tie their new methods to the ideology of utilitarianism. As we shall see in the next section, Bessey's famous *Botany for High Schools and Colleges* exemplified these features.[34]

The development of professionalism was, of course, based on foundations laid by the pre-Civil War scientific community. Historians have recently shown that American science grew rapidly in the 1840s and 1850s, with scientific curricula first introduced to the colleges, the first scientific schools established (Sheffield at Yale and Lawrence at Harvard), scientific societies organized, and the first journals established. The Museum of Comparative Zoology, founded by Louis Agassiz in 1857, for instance, pioneered in the method of graduate laboratory instruction (departing from the recitation method). Nevertheless, it was not until after the Civil War that a university oriented its entire graduate education around seminars and laboratory instruction. Johns Hopkins University was launched in 1876, on the model of the German university (albeit a misperceived and misunderstood model), under the guidance of Daniel Coit Gilman. The large midwestern public universities, led particularly by the University of Michigan under James Angell, paralleled the private eastern universities in expansion of the scientific laboratory. At the University of Wisconsin, under the presidencies of Charles Kendall Adams and Charles R. Van Hise in the 1890s and 1900s, the research ideal was tied to the notion of the laboratory (even in the social sciences). The laboratory became the link between the university as a service institution and the needs of the state.[35]

The laboratory stimulated professionalization of science in several ways, both sociological and intellectual. Sociologically, the laboratory made possible a new mode of teaching more directly related to training in technique than was the recitation method. Technique, or methodology, was—and remains—central to a professional discipline. The instrumentation of the laboratory allowed definition of mastery in a way that the recitation section, with the teacher taking the role of a quizmaster, could not. A simple indication of this is that before the mid-nineteenth century there were nearly no textbooks on how to observe nature for the purpose of collecting specimens, the traditional botanical task. Almost immediately with the development of the laboratory, the introduction of the microscope, and the use of apparatus like root pressure gauges, a flurry of textbooks and laboratory guides was published. The laboratory had, ironically, more of the spirit of the painting or sculpture classroom, where the students worked individually on common projects, helping each other, with the teacher circulating among them, offering criticism and guidance. At the hands of a wise and humane teacher, the laboratory would spontaneously ignite with excitement; under Bessey at Nebraska, the New Botany certainly did.

Similarly, the laboratory made possible a new intellectual framework for science: the specialty. Specialization in science has occurred philosophically and methodologically when a certain group of phenomena can be treated as distinct from the general subject matter of a discipline, but linked to the general subject matter by an agreed-on logic. This logic need not be deductive, in the sense that the central propositions of the specialty must be deducible from the fundamental theorems of the discipline, but can be procedural. The logic linking specialty to discipline can be the methodological steps by which a group of phenomena are experimentally analyzed and isolated from the general phenomena of the discipline. Thus, for example, before the phenomena of plant physiology could be *theoretically* linked to Darwin's theory of evolution, experimental plant physiology was a viable specialty, linked to observational biology by a set of *procedures* in which the functions of the plant in nature, necessary for survival, were experimentally analyzed in the laboratory. The intellectual framework of the specialty is the micro-paradigm. As we saw in the previous chapter and will deal at length with later, the microparadigm provides a research program for the specialists. Testable hypotheses are generated out of central assumptions, and consensually agreed-upon steps are taken in testing them in the laboratory.[36]

With his commitments to the laboratory method, physiology, tough-minded nominalism in science, and rigor in scientific logic, Bessey initially refused to accept the scientific credentials of the new specialty of ecology. He considered it a fad, dilettantism, and generally a waste of time in teaching and research. He perceived it as perilously close to "nature study." His reluctance to follow his distinguished students down their path to revitalized field studies provided a general milieu of skeptical doubt that personally spurred the students, particularly Clements, to quantify rigorously their field techniques. Bessey's role was crucial: he was the teacher of these men, revered for his personal inspiration and for his professional leadership. Sloppy science would not be acceptable.

The seriousness of Bessey's attitudes was perfectly expressed in his reaction to a suggestion that ecological field study be given first priority in high school botany. Such a recommendation was made by the Plant Morphology Committee on College Entrance Examinations of the National Education Association (NEA). The group was chaired by John Coulter, who handled the committee's business by correspondence—a method that did not lend itself to ironing out differences. Bessey called the recommendation a "pedagogical absurdity," and threatened to dissent from the committee's report if it persisted in the recommendation. "We must remember," lectured the Nebraska botanist, "that before we can [undecipherable] or teach of relations, we must know somewhat [sic] about *the things related.*"[37] Coulter, who was sponsoring ecological study at the University of Chicago and was the teacher of two important early ecologists, John Schaffner and H. C. Cowles, was not prepared to accept Bessey's wisdom. Coulter had taught ecology in high schools recently for a few years and was satisfied by the success of the method.

> I take it that the real fact is that any kind of botany may be taught successfully by a properly trained teacher. I do not believe that you are justified in speaking of this proposed method as a "pedagogical absurdity." It seems to be anything else to the professionals in pedagogical work, to whom it has been presented.[38]

The committee followed Coulter. Bessey dissented, as he had promised.[39]

Bessey was no less stringent with regard to research in ecology than with teaching ecology. A few years later, the same NEA committee again took up the question of what sort of botany ought to be taught in secondary schools. Bessey found—pleasantly to his surprise, no doubt—that he agreed with something in the report:

> That is a good hit which I find in the last line of the matter pertaining to Ecology in which the authors of the report deprecate "haziness and

guessing in Ecology." That is good. That is exactly what nine-tenths of the ecological stuff is now-a-days. It is "haziness and guessing." Of course, you understand that there is such a thing as good, solid work in Ecology, but there is mighty little of it being done.[40]

To the English physiologist and ecologist, A. G. Tansley, who would shortly become one of Clements's strongest defenders in Europe, Bessey called ecology "somewhat of a fad among a certain class of botanists."[41] When Frederic Clements's *Research Methods in Ecology* was published in 1905, it was praised for its rigor, for its strict attention to causality, and for putting ecology on a scientific basis.[42] These features were not simply idiosyncratic of Clements; they were characteristic of the Bessey environment.

Botany for High Schools and Colleges

For most advocates of the New Botany, the meaning of the laboratory method was exemplified by C. E. Bessey's *Botany for High Schools and Colleges*, published in 1880. It was hailed in reviews and private correspondence, and has since been historically evaluated as the single most important book published in American botany. It was immediately used by John Merle Coulter, teaching at Wabash College in 1881, along with Julius Sachs's *Lehrbuch* (Textbook of Botany), to which Bessey considered his *Botany* a good introduction, and Gray's *Manual*.[43]

At a superficial reading, the *Botany* appears only to be a laboratory manual, filled with exercises and diagrams, and connecting descriptive and explanatory text. This impression is a mistake, however. Bessey's text was not a book of recipes; it was not even simply intended to teach a science. It was intended to train the scientific mind. A student working through the book would be mentally disciplined with a critical method and an internalized philosophy of pragmatism. Successful completion of the text would make a scientist out of a student. In this deepest sense of creating an independent and critical mentality, able to investigate nature and form judgments without the active guidance of a teacher, the *Botany* provided a professionalizing education. Addressing the student who wished to make plant science a "special study," Bessey wrote, "this book aims to lead him to become himself an observer and investigator, and thus to obtain at first hand his knowledge of the anatomy and physiology of plants."[44]

Appropriately, the text did not begin with a statement of fact or with a "definition." It began with an instruction to act. The first line of the text is: "If we examine a thin slice of any growing part of a plant (Fig. 1) under a microscope of a moderately high power (400 to 500 diameters), there may be seen large numbers of cavities which are more or less filled with an almost transparent semi-fluid substance."[45] Bessey did not even begin by giving the student the proper name of what he or she was studying; rather, he indicated how whatever-it-was could be seen and what it would look like. The name was unimportant. The name was only a convenient label for shorthand discussion among scientists who knew how to interact with the phenomenon. The name appeared after fifteen lines of instruction and description of attributes: "protoplasm." This initial paragraph would have startled a botany student of the previous generation, who had begun his study of botany by memorization of definitions of species names.

The style of the *Botany* was no mere accident of the character of the book. The intention behind it was repeated in the other popular textbook that Bessey authored, *The Essentials of Botany*. There, Bessey twice warned teachers against requiring students to learn names and descriptive terms before they had obtained firsthand knowledge of plants by physical and microscopic examination. As did the *Botany for High Schools and Colleges*, so also did *Essentials of Botany* begin with an imperative: "Select a well-grown specimen of any plant."[46] In his discussion of protoplasm, the set of instructions on how to see protoplasm was followed by six tests to indicate whether what was seen was indeed protoplasm. These tests included staining and heat and chemical treatment.

Bessey's concept of protoplasm was, therefore, not an "idea." It was a staged sequence of activities, which resulted in agreement between scientists that they were talking about the same substance. It is important to recognize that for Bessey no scientific concept was private in meaning; scientific concepts were community concepts. Understanding what protoplasm meant included an ability to communicate with other scientists about the substance. The protoplasm discussion was constituted by three stages of action. First, the student examined a slice of plant under a microscope until a clear view was obtained of a substance having a described appearance. Second, the student applied a set of six tests to the substance to determine whether it was what other scientists referred to as protoplasm. The tests proceeded, for instance: "1. If a protoplasm mass is moistened with a solution of iodine, it at once assumes a deep yellow or brown color. 2. If treated with a solution of copper sulphate and afterwards with potash, it assumes a dark violet color. . . . 6. Protoplasm coagulates upon the application of heat (50 degrees Centigrade), or when immersed in alcohol or dilute mineral acids."[47] With assurance thereby obtained that the phenomenon was protoplasm, the student could proceed to the third stage, which was to conduct a series of experiments that would exhibit its characteristics as a biological substance and its role in the plant's life. These experiments established that protoplasm could "imbibe" water; move in characteristic ways, for example, stream; and have its movement suspended by external influences, for example, electric discharge.[48]

Reviewers of Bessey's *Botany* commonly pointed out that the book was an adaptation of Julius Sachs's *Lehrbuch*, and we may well suppose that Bessey's pragmatism was an Americanization of the German scientist's laboratory method. Bessey himself did not take pains to point out the uniqueness of his venture and even masked the difference between his and Sachs's work by informing the reader in the preface that "Part I follows pretty nearly that of Sachs's admirable 'Lehrbuch,' and in many instances it has seemed to me that I could not do better than adopt the particular treatment which a subject has received at the hands of the distinguished German botanist."[49] But Bessey's disarming remark is misleading; he did not in any way translate Sachs's book, and the underlying point of view about botany is significantly different. Both books were exemplars of the new laboratory botany, but Sachs's work had a subtle debt to romantic *naturphilosophie* and was distinctly Germanic in its expository style, while Bessey's *Botany* exuded nominalism and pragmatism. Bessey's *Botany* was American, not simply an Americanization.

Comparison of the two books immediately reveals the differences. Julius Sachs's 858-page treatise (in its 1875 translation, compared with the 611 pages of Bessey's *Botany*) began, not with a set of laboratory instructions but with a "preliminary inquiry into the nature of the cell," and is characterized by lawlike generalization, for example, "the law of configuration that prevails in all cells." How distant from the spirit of Bessey's approach! Sachs's preliminary inquiry derived from his assumption that the cell had a "nature" to be discussed, that it was more than a phenomenon, an aggregation of effects revealed when the experimenter interacted with it. This assumption of Sachs reflected his debt to German *naturphilosophie*, an idealistic interpretation of nature shared earlier in the century by Goethe, Schelling, von Baer, and Schleiden, for instance. *Naturphilosophie* rested uneasily with Sachs's Darwinism and exposition of laboratory empiricism, but appeared strongly in his discussion of plant growth. Here the great German botanist introduced the nineteenth-century concept of "original types," in explaining that the nearly infinite variety of plant parts and expression of plant members during growth could be reduced to a few original forms, or types. Sachs wrote definitively:

> Morphological investigation has led to the result that the infinite variety of the parts of plants, which in their mature state are adapted to altogether different functions, may nevertheless be referred to a few *original forms*, if regard is paid to their development, their mutual positions, the relative time of their formation, and their earliest states.[50]

He accepted one of the fundamental tenets of German philosophy in science, that the part cannot be understood without knowledge of the whole, that the individual stages of growth cannot be understood without knowing the complete history of the plant.

> The shape and size of the whole plant, the form, structure, and volume of cells are subject to regular changes, and their nature cannot therefore be inferred from the knowledge of one single phase, but rather from the sum of changes which may be called the life-history of the cell.[51]

The view that Sachs was thereby opposing may be called "reductive materialistic mechanism," the point of view that when the laws of the life mechanism in the smallest operative part of the plant are known, the life of the full plant can be predicted. Sachs was not asserting that the plant had a spiritual wholeness that influenced its growth, nor was he saying that the life of the plant as a whole was more than the aggregation of its parts; he was too late in the century for simple romantic holism in biology. He was, nevertheless, indebted to the romantic conceptualization of the problem of understanding the growth of the plant.

Sachs recognized that his discussion of plant growth echoed that of Goethe, who had taken an interest in science in the 1780s and 1790s, as part of his ministerial position in the duchal government of Saxe-am-Weimar. Goethe composed two scientific treatises, both vigorous refutations of Newtonian approaches to their topics, *The Metamorphosis of Plants* (1790) and *Theory of Color* (1805–10). In a footnote to a discussion of plant growth, Sachs wrote defensively, "It was these phenomena which first called Goethe's attention to the metamorphosis of leaves; at present the doctrine of metamorphosis rests on a better scientific foundation."[52] Goethe had argued that the entire growth of the plant was a series of metamorphotic transformations of the seed-leaf, and further, that seed-leaves could be categorized into a few "original types" or Ur-forms, as sharing essences. So for Sachs, many different parts of the plant were all "equally called leaves":

> For instance, the thick scales of a bulb, the cuticular appendages of many tubers, the parts of the calyx and corolla, the stamens and carpels, many tendrils and prickles, &c., are, in these respects altogether similar to the green organs which have been termed simply leaves (foliage-leaves). All these structures are therefore equally called leaves; and this designation is frequently justified by the fact that many of these organs, under peculiar conditions, actually become transformed into green leaves. Since the green organs which are termed leaves in popular language (foliage-leaves) may be considered as the primary form of leaves or as leaves *par excellence*, the remaining structures, which are also recognized as leaf-like, are termed changed, transformed, or metamorphosed leaves.[53]

In its embryonic growth, the parts of the plant were similar; frequently, it was impossible to distinguish them in terms of function. Only in more mature stages did they differentiate into adaptive functions, in the Darwinian sense of adaptation. Sachs considered the parts of the plant in the early stages of growth as leaf-form (except for the root, which was histologically and morphologically distinct).

> The older a member part of plant becomes, the more obvious becomes its adaptation to a definite function, the more completely is its morphological character often lost. In their earliest states the members to

which the same morphological names are applied (e.g. all the leaves of a plant) are extremely similar to one another; at a subsequent period all those distinctions arise which correspond to their different functions. With reference to these relationships we may now obtain a definition of Metamorphosis which can be used in a scientific manner:—*Metamorphosis is the varied development of members of the same morphological significance resulting from their adaptation to definite functions.*[54]

Bessey would have none of this. A comparison of discussions by the two scientists of the same topics immediately shows that Sachs's vestigal *naturphilosophie* was not carried into Bessey's *Botany*. Bessey went out of his way to indicate that in embryonic stages, plant members were indistinguishable, not because they were all forms of the primitive leaf, but because they were undifferentiated, looking like nothing in particular. In later stages, plant parts developed out of a *general* plant-body, not by transformation out of a leaf-form. Bessey's long refutation of metamorphosis is worth quoting:

> Every plant, in its earliest (embryonic) stages, is simple and member-less; and every member of any of the higher plants is at first indistinguishable from the rest of the plant-body; it is only in the later growth of any member that it becomes distinct; in other words, every member is a modification of, and development from, the general plant-body. Likewise, where equivalent members have a different particular form or function, it is only in the later stages of growth that the differences appear. All equivalent members are alike in their earlier stages, whether, for example, they eventually become broad green surfaces (foliage leaves), bracts, scales, floral envelopes, or the essential organs of the flower.[55]

The differences in meaning between Bessey's discussion and Sachs's were not the result of any difficulty of translating into English. Sachs's language was tied to the tradition going back to Karl Ernst von Baer and Goethe, while Bessey's language lacked the references to metamorphosis, original forms, or identity of members that would mark allegiance to *naturphilosophie*. Sachs believed that the common manner in which all leaves grew out of stems and branches could be conceptualized independently of the physiology of the leaves:

> But if the one characteristic only is kept in view that they all bear leaves which arise below their growing apices, an agreement is found as important as it is complete, which may for the time be altogether

abstracted from the physiological functions and the corresponding structure.[56]

I have not discovered a passage in any of Bessey's writings in which he wrote that biological functions of plants can be abstracted. Scientific statements in botany were always tied to actual plants or translatable into statements about methods of examining actual plants. Bessey was a nominalist.

Bessey's philosophical approach to botany, exemplified in his landmark textbook, may properly be labeled pragmatic. As such, it bore close resemblance to the pragmatic style of thought that was emerging in the 1880s in many areas of American intellectual life: the "new jurisprudence," the "new social sciences," the "new engineering," and the "new medicine." Pragmatism was tied primarily to experience and only secondarily to deductive formalism and logic. His own generation believed they had broken away from the methodology and wisdom of learning of the previous generation. Bessey's war cry, "Botany is the study of *plants*, not the study of *books*," was an analogy to the motto of the "new" learning of the 1880s.[57] Before that decade, professions tended generally to be taught by apprenticeship and rote learning, in which memorization and deductive inference from general principles were important. In the new era, professional education was taught in the laboratory, or its equivalent, such as the moot court, grounded in the experimental method, with ideas linked by function rather than logic.[58]

We may see the peculiarly American quality of Bessey's thought by turning to Charles Saunders Peirce's classic essays, "Fixation of Belief" and "How to Make Our Ideas Clear," which appeared just before Bessey's *Botany*. Charles Peirce (1839–1914) was a logician and philosopher, and one of the founders of pragmatism, although he later tried to distinguish his views from the pragmatism of William James and John Dewey. By his own account, he was trained as a chemist and was greatly influenced by discussions with Harvard philosopher, Chauncey Wright, but he was never able to obtain an academic appointment and passed his years working in the U.S. Coast Survey.[59] Peirce's pragmatism was an adaptation to philosophy of what he perceived as the implications of Darwin's theory of evolution. Peirce treated ideas, in this Darwinian view, as instruments in the struggle for survival, and the goodness or badness of ideas was therefore, in his view, a matter of whether they established habits of action that contributed to survival.* From this general perspective, the difference between ideas reduced to the difference in their effects, that is, the difference they make in our survival. This brief expression of Peirce's early philosophy, of course, does not do it justice, but it does orient us to see the role of "testing" in distinguishing between ideas and in relating ideas or conceptions to reality. His notion of testing particularly will bring us to Bessey's conception of the scientific method.

For Peirce, the purpose of thought was to establish belief, that is, a predisposition to act or a habit of acting in particular ways. In the Darwinian philosophy, these habits would have to contribute to survival. Thought, or inquiry, moves from doubt, arising by effective challenge to a habit, to belief, or "settlement of opinion," when we feel confident in automatically responding to a situation. In "The Fixation of Belief" (1877), Peirce reviewed various methods of establishing belief, such as reliance upon authority, only to reject them as of limited usefulness. The scientific method was for him not a method simply of discovery of "truth"; it was the only generally reliable method of settling opinion. The scientific method was distinguished from some other methods by the capacity to settle opinions in a community, whereas mysticism was confined to private validity. It was, then, a method for establishing a consensus among individuals who might previously have disagreed.

The public character of the scientific method appears in Peirce's discussion of the meaningfulness of ideas. To be meaningful, a scientific idea must be clearly distinguished from other ideas, and these distinctions accepted consensually by the scientific community. What could distinguish scientific ideas? Nothing else than the effects they produce. What distinguishes the idea of protoplasm from the idea of gelatin?—That the ideas lead the scientist to different courses of action. If the scientist believes that the substance in his laboratory is protoplasm, then he stains it, heats it, and acts on it in procedures that will tell him if this substance behaves in ways

---

*I have avoided using the term *validity* when referring to the characteristic of ideas that was in Peirce's view related to the struggle for existence. Peirce tended to use validity, goodness, and meaningfulness of ideas interchangeably, whereas in modern logic, validity is reserved for the characteristic of propositions relating to the rules of inference, while meaningfulness related to what Peirce called matters of fact, or what we call contingent propositions.

other scientists assure him protoplasm will behave. If the scientist believes that the substance in his laboratory is gelatin, he removes it to his kitchen and prepares his cooking with it; if fruit pies set according to the recipes, then he is assured he has gelatin, not protoplasm. As Peirce said, "to develop its [an idea's] meaning, we have, therefore, simply to determine what habits it produces, for what a thing means is simply what habits it involves."[60] Put differently, "our idea of anything *is* our idea of its sensible effects."[61] Protoplasm stains a certain color and does not jell pies; gelatin does not stain the same color as protoplasm and does jell pies. If protoplasm and gelatin did not differ at all in their effects, it would not be scientifically meaningful to assert they are different substances or that our ideas of them are different.

Peirce's thesis was the same that underlies Bessey's famous *Botany* (1880). The scientific method, then, was a method for establishing consensus among scientists on the meaning of scientific ideas, or propositions, by experimentally determining what were the sensible effects that distinguished one idea from another. Science was a procedural set of actions conducted with a community of scientists as participants and audience. It was not metaphysical; it was antimetaphysical. The metaphysical abstractions that insinuated themselves into Julius Sachs's *Lehrbuch* would not be present in Bessey's rendition of that pioneering German text. I do not mean to imply by this discussion that Bessey obtained his vision of what science was to be from reading Charles Peirce's classic essays. I do not know whether Bessey did read them, though there is little reason to doubt that the Nebraskan scientist did read the *Popular Science Monthly*, then a respectable scientific journal. Bessey did not have to. He was responding to the same Darwinian influences, trained in the same laboratory methods, as Peirce. Bessey embodied the pragmatic method about which Peirce wrote, and transmitted it to his students who later established ecology.

3
QUADRAT:
Quantitative Methods for Ecology

Quantification in Plant Geography

In 1897, Roscoe Pound and Frederic Clements took a daring leap into numerical quantification of ecology. Their leap involved a qualitatively new way of treating ecology, in which it was assumed that the primary sensory experience of the botanist was misleading and that only exhaustive measurement could elicit the true relations among plants. It is tempting, on a superficial reading of the botanical literature, to regard Pound's and Clements's quantification of plant geography in *The Phytogeography of Nebraska* as a simple extension of early traditions of quantitative floristics, which is how Edouard Rübel, the Swiss botanist and constant critic of Frederic Clements, regarded their work in a review article in 1920.[1] Such a reading is in error. Neither of the two traditions of quantification in nineteenth-century botany which appeared to lead to Pound and Clements could have been extended logically to their achievement. Quantitative floristics, begun by Alexander von Humboldt, had run into a sterile dead end along a course involving the quantification of sensory experience. This tradition I characterize as "qualitative quantification." Quantitative anatomy and morphology of plants, the second tradition, was making little progress in the 1890s, unable to integrate itself with statistical theory, which was being developed in England and the United States by biometricians. Pound and Clements had to deny the fundamental assumptions of traditional quantitative floristics, reformulate plant geography on a new conceptual base, and devise a new methodology for field investigation. In so doing, they tore ecology away from its past and thrust it into the twentieth century.

Plant geography originated as a formal study in Europe in the early nineteenth century. Botanical exploration of the Americas and the South Pacific in the previous century vastly augmented European knowledge of the floras and environmental conditions in semitropical and tropical climates. The travels of Captain Cook, with his companions Reinhold Förster and his son, George Förster, to the South Pacific stimulated Alexander von Humboldt to undertake foreign exploration, ultimately leading him between 1709 and 1804 in the first scientific expedition in Spanish America in nearly two generations. This exciting and perilous journey, reported in exquisite detail in the multivolume *Personal Narrative*, led to Humboldt's *Essai sur la géographie des plantes* (1807), considered the founding treatise in plant geography. Within a few years, the Swiss botanist, Augustin de Candolle, friend and correspondent of Humboldt, joined the ranks of the pioneers in the new specialty with his "Essai élémentaire de géographie botanique" (1820) and more specialized monographs treating the phytogeography of France and the valley of Léman, Switzerland. By the decade of the 1830s, plant geography was sufficiently developed to have practitioners in the major European countries and divergent philosophical viewpoints about the enterprise itself.[2]

Humboldt took the first major step in quantification in plant geography. Though not unanticipated in the botanical concerns of the eighteenth century, and drawing upon developments in cartography related to natural resource surveys, Humboldt nonetheless boldly established a strict mathematical framework for analyzing the geographical distribution of species of plants. Native plants were suited to their climatic environment. The distribution of species and genera ought to show patterns related to distance from the equator, since temperature declined as one moved northward or southward from the equator. These planetary patterns of distribution allowed comparison of geographical areas in terms of the ratios of genera locally represented to the total number of genera, or of the percentage of species native to an area to the total number of species in their genus. These ratios could be roughly correlated to climatic indexes, such as temperature, frost-free days, and rainfall.

In later decades, Humboldt invented other quantitative, cartographic devices for the analysis of plant distribution. One of the most useful was the *isotherm*. The isotherm was a line drawn on a map connecting points of equal, annual, mean temperature. It provided a precise definition of climatic area as a basis for examining ratios of genera and species. Without statistical techniques, such as correlation, it was difficult to compare the congruence between a climatic area contoured by isothermal lines and the distributional area of a species or species-mix defined by ratios; but laid out on a map, the floral lines and the isothermal lines did provide the first technique for analyzing the relationship between habitat and plant type.[3]

Humboldt's understanding of plant distribution was based on his conception of the relationship between plant types and climatic areas. He believed that every category of climate had one particular form of vegetation ideal for it. Some climates ideally nourished forests, others grasslands, or desert, or alpine flora. Hence, Humboldt wrote frequently about the "physiognomy" of the landscape, or literally, its "face." The face of a landscape was unique, as are all human faces, but at the same time it expressed the ideal qualities of that climatic region, just as human faces exemplify ideal racial features. Humboldt's faith in the Ideal Type underlay his mathematicization of botany. The congruence between vegetational form and area ought to be mathematically identifiable as one-to-one. Humboldt's typological thinking was shared later in the century and underlay efforts of botanists to describe the "formation," the vegetational unit having its own unity, just as a human face, comprised of individual parts, composes a whole.[4]

After Humboldt, quantification in geographical botany took two directions. One followed Humboldt's insights, utilizing isothermal lines and species-ratios (floristics) to define the climatological and geographical boundaries of species. Another, dating from mid-century, laid down an arbitrary surveying grid on a large geographical area: each large section within the grid was of equal area and called a *quadrat* (or *quadrille*, or *quadrate*). Quadrats could be compared according to which species were represented in each. The former generated a natural area containing a species, the latter an artificial area. The practical advantage of the quadrat method was that the sections were fixed and easily determinable, thereby aiding the counting of species. The method of natural geographical areas was useful on a continental scale, but at the regional level, for example Great Britain or the Rhine valley, the isolines were difficult to fix, since local climatic conditions varied from year to year. On a continental scale, changes stabilized about a mean.[5]

Progress in the utilization of natural climatological and geographical boundaries, on the one hand, tended to be in terms of refinement of isothermal lines. For example, an isotherm of annual mean temperature was quickly found to be inaccurate and did not correlate, even to a casual examination of a map, to characteristics of vegetational distribution. Consequently, lines for annual mean high temperature, annual mean low temperature, date of first autumn killing frost, and date of last spring killing frost, for instance, were devised. This work ultimately reached a dead end, because there were no equally exact methods for the quantification of individual plants in distribution to be correlated to the climatological districts. Also, the statistical techniques for analysis of correlation were not invented until the 1880s and were not widely used for several decades after.[6]

The quadrat, on the other hand, was increasingly and successfully used to obtain indexes of frequency, that is, of the number of stations or locales at which a species was found within its general distributional area. The quadrat was first successfully used in 1879 by Heinrich K. H. Hoffmann in his study of the middle-Rhine valley. A grid of forty-nine quadrats, each 21.4 kilometers square, was superimposed on the valley. Frequency was measured as the number of quadrats in which a species was located in at least one locale; for instance, *Berula angustifolia* was found in fourteen quadrats, *Biscutella laevigata* in five. *Berula* was more widely distributed within its geographical area than *Biscutella*.[7] In the 1890s, Oscar Drude, the German follower of August Grisebach, who was one of Humboldt's chief disciples, used the quadrat method in his investigation of the plant geography of Germany. Floral territories were established to an exactitude equal to the size of the quadrat.[8]

Useful as the quadrat was, the large size of the grid prohibited complete precision, that is, exactitude at the number of individual plants. The traditional term for the number of individual plants in an area was *abundance*. As late as 1893, abundance was pronounced incapable of quantification. John Briquet, in his article on mathematical botany of that year, stated firmly, "Le degré d'abondance des espèces ne peut être indiqué que part des expressions plus ou moins vagues." Rather than counting individual plants, botanists developed ranking schemes, in which sensory impressions of abundance were assigned to grades. The botanist would survey a natural scene and comparatively rank impressions of the number of individuals of species in this locale against each other or against the same species at another locale. Efforts were made to quantify the number of individuals, short of actually counting all of them. Thus the Swedish botanists, R. Hult and R. Sernander, developed in the 1880s and 1890s a density-ranking scale for vegetational cover involving ratios of species and the distance between plants of a species. These quantifications remained, however, tied to levels of prior sensory impressions. All botanists at this time with whom I am familiar thought that abundance would remain outside the exactitude of a progressive science.[9]

Drude's discussion of abundance in *Deutschlands Pflanzengeographie* (1896) revealed the qualitative character of quantification in plant geography at the moment that Pound and Clements jumped to fully numerical quantification. Drude distinguished between five grades of abundance: social, gregarious, copious, sparse, and scarce (in order of decreasing abundance). A species that was "social" appeared in such numbers that the scientist could not be concerned with absolute numbers of individuals. A species of grass dominating a meadow would be a social plant, as would a moss populating a stump. A gregarious species of plant appeared to the botanist as "herds" of plants in the midst of other species. The copious plant appeared singly, but plentifully, within the general abundance of other plants. Modern statisticians recognize these distinctions as ordinal rankings, that is, without intermediate values. The categories were based on years of accumulated field experience by botanists and reflected qualitative differences to the unaided, though not untrained, eye. The sensory perspective is easily discoverable. The botanist surveys from a hill a nearby forest. The individual trees cannot be counted, but differences in form, color, and size permit the botanist to identify species and to establish grades of abundance. In the eastern Canadian broad-leafed forest, the most abundant tree would be the "gregarious" sugar maple. In a rapid survey, the botanist would overlook the black cherry. If called to his attention by a more careful observer, the botanist would grade the cherry as "sparse" for this station.[10]

The qualitative quantification of abundance would work rather well for the European setting in which it was developed. The terrain changes in brief distances, often dramatically. Within small geographical areas several formations are often present: marsh, meadow, and forest, for instance. The rudimentary categories of abundance reflected the abrupt and primitive distinctions in vegetation presented to the eye. With this background in mind, it is obvious that these categories of abundance could not easily be applied to dissimilar terrain. Imagine the same botanist trying to grade the abundance of species while standing on a rolling hill by the Missouri River in eastern Nebraska, surveying westward the limitless Great Plains. When the scientists at Nebraska tried to apply Drude's categories, they failed. Out of that failure, they made the leap to a numerically quantified ecology, based on the interval variable.

Statistics in Plant Anatomy and Physiology

The use of statistics in describing the anatomical features and functions of the individual plant was more familiar to the Nebraska scientists than German quantitative floristics. As the premiere American plant physiologist, Charles Bessey employed statistics in this way. But even this tradition of mathematization was not without its problems. For with the possible exception of Bessey, this use of statistics was largely descriptive and frequently incidental to the argumentative structure of the botanical treatises. Bessey was one of the few scientists to conceive his scientific argument in statistical terms. Significantly enough, he was undertaking this work, regarding especially plant evolution, at the moment that Pound and Clements were trying to figure out what to do with their failure in applying Drude's analytical framework to Nebraska plant geography.

A few examples drawn from the *Botanical Gazette* in the 1890s will illustrate the relatively unsophisticated statistical analysis. The *Botanical Gazette* was founded and edited by John Merle Coulter, who had built a reputation equaling Bessey's, while traveling from Wabash College to the presidency of Indiana University, to Lake Forest College, and finally, in 1897, to the University of Chicago, where he was head of the Department of Botany and presided over the prestigious facility made possible by the Hull endowment.[11] The journal appeared quarterly, carried intimate news of the profession, and was easily the most advanced botanical journal in the United States, moving away from systematics into physiology and ecology before the other major journals took this trend. It was undoubtedly widely read by the Nebraska group, since they took the trouble to forward to its "News" column personal notes about their activities, and occasionally published articles in it. *Botanical Gazette* therefore tells us of the current state of the discipline in the Midwest.

The most common use of statistics was the simple dimensional measurement of mature plants, for example, length and breadth of leaves and distance between nodes on the stem. Seldom were dimensions given as means or a range and variance specified. The lack of any analysis of variations was the more obvious when it was treated, as in an article by M. E. Meads, "The Range of Variation of Species of Erythronium" (1893). Meads carefully measured every part of the flowers of *E. americanum* and *E. albidum*, enabling him to generate sufficient data for tables which included mean, number of plants above the mean, and number of plants below the mean. The table was more sophisticated than the raw presentation of data found in most other data tables in botanical literature, but nevertheless it was severely limited. It did not, for instance, provide the simple but necessary information we obtain from the standard deviation. Because Meads's article was directed toward range of variation, the failure to utilize the new statistical techniques developed in Belgium by Quetelet and in England by Galton and Pearson for variation in human populations is somewhat puzzling, since these statistical techniques were developed precisely to deal with the problems of variation in the context of Darwinian theory in which Meads was working in botany.[12]

In other publications, numerical data were accumulated in discussing the time of movement of granules in protoplasmic streaming and the extent of water transpiration, the abundance of bacteria at different depths of the Atlantic Ocean off Woods Hole, the angle of leaves at times of day, together with temperatures, and the measurement of plant root cells in order to determine the causes of the curvature of roots.[13] None of the publications concerning these researches utilized the correlation, the normal frequency distribution curve, or variation or standard deviation, although all of these statistical concepts existed then and their use would have been important in the inferential structure of the arguments. For instance, S. G. Wright, who conducted research on leaf angle and movement, provided a graph showing two changing variables, leaf movement (change in angle), and temperature change. He implied that these two variables changed in relation to one another, but he did not provide correlational analysis, recently invented by Francis Galton and developed by Karl Pearson, which would have allowed him to analyze directly the relationship between the variables.[14] Similarly, H. L. Russell did not use correlation regarding the relationship between abundance of bacterial flora and depth of the ocean, though his table of data is presented in such a way as to imply strongly that such a relationship was present.[15]

My discussion of the inadequacies of statistical analysis in these botanical researches is not a criticism of them. The publications were in the forefront of the application of statistical data and sophisticated analysis was soon widely used in botany. The shortcomings show how the data generated in plant anatomy and physiology tended to remain outside of the logical structure of the argument presented in the publication, that is, to remain descriptive. Quantitative botany was not yet incorporated in terms of quantitative problems. The data were not central to the argument, because the argument was not basically about numbers, but about something else.

Biologists in the years from 1896 to 1899 were generally aware of their need to master the new techniques. Charles Davenport, a zoologist at Harvard University, who later became famous for his ideological advocacy of eugenics, stated in the preface to his pioneering American statistical text that the book was a response "to a repeated call for a simple presentation of the new statistical methods in their application to biology."[16] In writing this text, he drew upon the studies of variability by Karl Pearson and E. T. Brewster.[17] The isolation of botany in the midst of these external changes in mathematical technique showed again that the move to numerical quantification by Pound and Clements was not a simple extension of earlier statistical traditions. The botanists had first to perceive the logical structure of the botanical thesis in a new way, then they could move to the new techniques.

This situation is thrown vividly into relief when we turn to one of the rare researches that was quantitative in the structure of its logic: which recognized the problem as quantitative in definition and the object under examination as statistical in character (rather than simply being an object capable of statistical description). Such was Bessey's "Phylogeny and Taxonomy of the Angiosperms" (1897), in which Bessey argued for a particular evolutionary path for the present day Angiosperms (true flowering plants) from paleontological data.[18] His data were the percentages of total plant fossils in each paleontological period comprised by certain orders of the fossil Angiosperms. The use of percentages immediately signaled that a statistical analysis was underway, since they were relative data, rather than raw data. These percentages were arranged in tables that showed the increase and decrease of percentages for each order of Angiosperms throughout the geological epochs up to the present. These changes were analyzed pictorially or quasi-geometrically by representing each percentage as a shaded portion of differently sized triangles (size dependent on the absolute number of fossils for that epoch). These triangles were organized into alternative family trees, which allowed Bessey to show the path of evolution, on the Darwinian principle that species more advantaged than others reproduce more prolifically, that is, increase their representation in the population of interbreeding individuals.

No single component of Bessey's argument or use of statistics was original. The use of the percentage of a group of species to the total number of species of an order to show geographical distribution had been developed by Humboldt ninety years earlier. It was an easy adaptation to shift these changing percentages from geographical locations to geological strata locations. In fact this adaptation had been made in the 1830s and was implicit in Charles Lyell's analysis of the paleontological series. And changing distributional statistics were widely used in floristics. Bessey's originality lay in the point of view that the debate over the evolution of Angiosperms was about *populations* of plants (rather than types of plants) evolving through geological periods, and could be conducted by statistical analysis without descriptive reference to individual plants. The thesis showed that Bessey had grasped the true meaning of Darwin's theory of evolution, since Darwin originated the notion in biology that populations of individuals evolved, rather than individuals themselves, though most of Darwin's contemporaries did not appreciate this innovation. H. C. Cowles remarked on its infrequent use in an article in 1901.[19]

Bessey's "Angiosperms" also reveals that geology remained a potent source of inspiration for statistics in botany. Glancing over the nineteenth-century traditions of quantitative botany, the role of geology was clear enough. Humboldt was originally trained as a geologist and his first important position was inspector of mines. The concept of the "formation" of plants in botany was apparently taken by Grisebach from geology in the 1840s, since the formation was a fundamental concept of historical geology. In the 1890s, we have not only Bessey's work in geology, but also that of H. C. Cowles, who was initially a student of geology at the University of Chicago, before he was lured into botany by John Coulter. This, of course, does not exhaust the list of botanists influenced by geology, but it does line up important contributors to plant geography and ecology.

The Leap to Numerical Quantification in Ecology

The appearance of Bessey's paper on Angiosperms in 1897 was an important tip-off to what was going on in the botanical group at Nebraska. Statistical examination of the flora, using percentages and population definitions of vegetational groups, had been under way at least since 1896. The next year, Pound and Clements developed their method of counting secondary species to generate more precise statistics to correct Drude's *Plant Geography*. The work of Bessey, Pound, and Clements in true statistical botany in 1896, if not earlier, revealed that a major problem shift had occurred which made such statistics possible and cast the problems of grassland ecology into statistical terms. What was this problem shift and how did it lead to numerical quantification in ecology?

The record book of the Sem. Bot. provides us with a remarkable chronology of the intellectual interests of the Bessey group, with Pound and Clements as the central student figures, during the 1890s. Together with other manuscript evidence, we can date with considerable accuracy the intellectual shift that led to dynamic ecology. In this log, the year 1896 is crucial. Before it, the Nebraska botanists had two straightforward concerns, plant physiology and Nebraska floristics. We have already seen that both of these interests revolved around Bessey, as is clear from the student theses and his own writings. The Botanical Survey of Nebraska originated in the minds of J. G. Smith and Roscoe Pound while the two were on a collecting expedition in July, 1892, and was formally organized as part of the Sem. Bot. the following month.[20] By the fall, the first publication of the survey was at the printer.[21] In the next four years, the record book of this enterprising group of botanists concerned itself with the expeditions, collections, and herbaria of the survey and the monthly meetings at which papers, largely on plant physiology, were presented.

At the end of the academic year in June 1896, the first indication appeared that the interests of some seminarians were changing. On

June 5, Roscoe Pound presented an oral report on "The Phytogeography of the Little-Blue Valley."[22] Pound had become acquainted with Oscar Drude's *Deutschlands Pflanzengeographie*. Bessey, as botany editor for the *American Naturalist*, had been sent Drude's book for review, and turned it over to Pound. In a reminiscence, Pound recalled reading Drude's book "in 1895," though this is probably either a mistake of the elderly man's memory or a typographical error in the publication.[23] In any event, the review was published in the June (1896) issue of the journal. Pound told Clements about Drude's book, and Clements responded enthusiastically.[24] In the fall, Clements followed his colleague's steps in phytogeography with a report to the seminar on "The Plant Formation as an Element" (October 10).[25] In the following month, an entire afternoon was given over to the exciting new field. The symposium was hosted by "Fred" Clements, and the topics served as general introductions to an audience presumably not well acquainted with the new specialty. Clements spoke on "Phytogeography, What It Is"; Bessey was given the senior scientist's role of evaluating "The Significance of Phytogeography"; Pound introduced "The Vegetative Covering and Its Subdivision"; Clements concluded with a practical application, "The Phytogeography of Nebraska."[26]

At some time during the summer or early fall (1896), Pound had suggested to Clements the idea of doing a phytogeography of Nebraska, apparently on the model of Drude's phytogeography of Germany.[27] It is possible, and indeed in line with the aims of the botanical survey initiated by the Sem. Bot. in 1892, that the suggestion for a geographical survey of the flora of Nebraska was made earlier, before the receipt of Drude's book in Lincoln. Clements reminisced much later that "by 1895, it had become evident that a detailed account of the vegetation of Nebraska would be of service to all the prairie states."[28] Nevertheless, it seems clear from the manuscripts, and more than abundantly clear from the internal content of *The Phytogeography of Nebraska*, that Drude's book formalized the general aims of the previously existing botanical survey with a specific theoretical aim. Pound and Clements recalled independently that the writing of the *Phytogeography of Nebraska* was begun in the autumn of 1896.[29]

The composition of the work proceeded rapidly. Pound and Clements wrote at night, in the university herbarium, just a few

doors down from Bessey's office. Since campus regulations prohibited any staff member below professor from using the buildings in the
evening, the two botanists entered and left the herbarium through a
window. Both Bessey and the night watchmen were aware of this
"patent subterfuge," but did not interfere.[30] They had to work in the
evenings, because Pound had an active law practice and Clements
was Bessey's assistant. As Clements pleasantly remembered:

> The two friends alternated in the major tasks of dictating and trans-
> cribing, pausing now and then to discuss a point, seek new inspira-
> tion, or to relax by whistling in unison snatches from favorite grand
> operas. At such times, Dr. Bessey would occasionally look in, to say
> that he knew the work was going well when the strains floated down
> the corridor.[31]

During the year, the two scientists reported to their colleagues in the
Sem. Bot. on the progress of their work. On December 5, 1896,
Pound gave a paper on mycological statistics, and on December 26,
another paper on "Outlines of 'Phytogeography of Nebraska.' " On
March 27, 1897, Pound gave another paper that utilized statistics,
this one concerning "Ecological and Distributional Statistics of
Nebraska Grasses." At the end of May (29), Clements spoke on the
plant formation and Pound outlined the habitat groups of the
prairies.[32]

Pound remembers that the writing took two years, but almost
certainly that is not correct without qualification.[33] The men intended to submit the *Phytogeography* as their joint doctoral thesis. Since
Pound received his doctorate (the University of Nebraska's first
awarded doctorate) in June 1897, I must assume that the work was
sufficiently completed to be accepted by Bessey. Clements's doctorate was held up a year, while he completed a minor requirement in
Spanish; the degree was awarded in June 1898.[34]

During the writing of this "thesis version" of the *Phytogeography*,
the two scientists had not developed the meter-plot or quadrat
method.[35] There is no indication whether they might have begun to
think about a new approach to determining abundance of species on
the prairies during the fall of 1896, or winter of 1897. Clements
recalled twice that the quadrat method was devised by them during
1897.[36] Furthermore, there was no mention of the new method in the
Record Book of the Sem. Bot., until October 16, 1897, when Pound
and Clements jointly delivered a paper to the group on a method of

determining abundance, almost certainly the substance of an article they published on meter-plot method in 1898.[37] Since it was impossible for the scientists to develop the method until the snows had left the prairies, we may say with some assurance that the quadrat method was worked out between the spring and late summer of 1897. Therefore, the meter-plot was invented after the scientists had made a thorough and systematic attempt, during the previous fall (1896) and winter (1897), to apply Drude's phytogeographical framework to the Nebraska prairies.[38]

We need to ask a series of questions about this invention of the meter-plot (the meter-plot itself I discuss in detail in the next section). Why should Drude's work on the phytogeography of Germany have made, in general, such an impact on the botanists under Bessey? Why should the application of Drude's framework of phytogeography to Nebraskan vegetation have led specifically to the invention of the meter-plot, as a symptom of a new way of looking at ecology? Finally, why should this shift have occurred in between 1896 and 1897, rather than earlier or later? We can understand the impact of Drude's book by turning to the floristic studies of Nebraska that the Sem. Bot. was undertaking in the years between 1890 and 1895 and by Bessey's work in floristics. Generally the empirical research the group undertook was similar to the empirical research undertaken by Drude, but lacked the theoretical framework of the German research. Rather, the Nebraska group's work had been applied research, tied closely in Bessey, and somewhat more loosely in his students, to the agricultural needs of the booming state.

After Bessey moved to Nebraska in 1884, he became interested in the geographical distribution of plants of that state.* This interest did not derive from an abstract curiosity about the problem, but directly from his responsibilities as botanist of the Agricultural Experiment Station, and hence his obligations to the farmers of Nebraska. The agricultural cultivation of the Nebraska prairie in the 1870s and 1880s permitted the intrusion of new weeds in soil from which the tough prairie sod had previously barred them. The magnitude of ecological change, and hence of Bessey's considerations of the

---

*The material in the following section is drawn from a previous publication of mine, "Theoretical Science and Technology in American Ecology," *Technology and Culture* 17 (October, 1976):718–28.

invasion of weeds, is illustrated in the increase of improved farm acres. In 1880, 5,500,000 acres were improved; by 1900, this acreage had more than trebled to 18,400,000 acres. The episode of the Russian thistle illustrates the way in which Bessey's interest in geographical distribution of plants was provoked. The Russian thistle, a bushy weed accidentally imported from Europe, established itself in South Dakota in the 1880s and spread to Nebraska in the early 1890s. The thistle was sufficiently severe to force abandonment of farmland in South Dakota. The spines could disable a horse and, if well rooted, the plant could stop a mechanical harvester. Bessey went to the fields to study the plant, involved the Botanical Seminar in the emergency, and recommended to farmers ways to halt the infestation. In the era before effective chemical control of pests, measures of "applied ecology" were often the farmers' only resort. Bessey consequently studied the life cycle of weeds, the means and speed of their geographical distribution, their competition with established crops, and other matters likely to be of interest in combating weed invasions.

The Russian thistles (tumbleweeds) were not the end of the farmers' problems, and in an annual report to the regents of the University of Nebraska, Bessey not only enumerated the afflictions of the state's agriculture, but pointed inadvertently to the wide geographical demands that the state's botanist could be exercised to meet. The farmers were faced with Asparagus Rust, a fungus that had appeared in New Jersey several years previous and was spreading westward. Carnation Rust, another eastern plant disease, had already been spotted twice in university greenhouses. Downy Mildew had struck potatoes, Stinking Smut had infected the wheat, and Loose Smut had afflicted wheat, oats, and barley. In western Nebraska, a poisonous plant (a wild species of larkspur) was spreading into grazing land and poisoning domestic animals. A variety of wild mustard had gained a footing in the northern counties of the state and was threatening pastures.[39] While this botanical firefighting may not have addressed basic scientific problems, it brought Bessey and his botanists continually against the relationship between the climatic and soil regions of the state and the infestation of undesirable plants.

Bessey's obligations to his farmer constituency also led him to study the geographical adaptation of foreign crops to Nebraskan soil.

He regularly reported to the state's horticultural and agricultural societies about problems of forage, the disappearance of the buffalo grass, and the reintroduction of forests. He published numerous articles on plant geography, which I have cited elsewhere, but which included among their topics the geographical distribution of pine trees, ironwood trees, the box-elder, the barberry, the black walnut, and the "most notable weeds of Nebraska." None of these publications was scientifically remarkable. They were not generally even addressed to basic scientific problems. But they did provide the general framework of applied research within which Bessey's students undertook floristic surveys of the state. The students made their surveys with assurance of the applied value of their efforts, regardless of the usefulness of the plants enumerated. The herbarium near Bessey's office grew yearly as the plant inventory of the state was elaborated. Therefore, it was not a totally unanticipated step when, in 1892, Roscoe Pound and J. G. Smith thought up the idea of a botanical survey of the state. A survey would formalize the nonsystematic surveys that Bessey and others had been making since 1884.

The important feature, which the botanical survey of 1892 had as a result of its natural growth out of the Bessey group's tradition of plant surveys, was its conception as an inventory, with knowledge applicable to the practical agricultural problems of the state. It was not conceived within a theoretical framework. In other words, the botanical geography of the Sem. Bot. in the four years preceding the receipt in Lincoln of Drude's *Deutschlands Pflanzengeographie* was not disciplinary in character; it was not in the tradition of plant geography from Humboldt, Grisebach, and Drude. The botanical survey was not intended to test hypotheses, or extend a paradigm, or solve disciplinary puzzles. From my reading of the publications of the group and from the manuscripts, I have no evidence that they were even aware of a theory of geographical distribution at this time. It is also possible that Bessey was unacquainted with the European literature on distribution. He was a physiologist and a laboratory scientist in disciplinary allegiance, not a floristic botanist.[40]

In 1898, to illustrate the departmental emphases, the library of the Department of Botany held twenty-five-hundred books and subscribed to forty botanical journals. Yet, it was not until 1899, or later, that many basic texts in plant geography were ordered for the botanical or campus general libraries, such as Humboldt's *De distributione geographica plantarum* (1817), and *Essai sur la géographie des plantes* (1807), Augustin de Candolle's *Regni vegetabilis systema naturale* (1818–21), August Grisebach's *Bericht über die Leistungen in der Pflanzengeographie* (1843–53), and J. F. Schouw's *Grundzüge einer allgemeinen Pflanzengeographie . . . Aus dem Dänischen übersetzt* (1823).[41]

If one reason that Drude's *Deutschlands Pflanzengeographie* of 1896 made such an impression on the Nebraskans is that it was their first taste of disciplinary geographical botany, based on the long European theoretical tradition, the question remains why that impression was made with this book and in 1896. Part of the answer, of course, is that without acquaintance with the older European literature, they would not have known of Drude's earlier work, published in 1890, which provided a schematic outline for plant geography of the world. But a more important part of the answer is that in the years 1890 to the mid-1890s, the group at Nebraska was not sociologically ready and their botanical survey had not gone far enough to demand a theoretical structure. Recall that the Sem. Bot. itself had declined in membership in 1890, resulting in the formal initiation procedures in 1892 for undergraduate botanists. Frederic Clements, who replaced H. J. Webber as Roscoe Pound's closest botanical friend, was only a sophomore in 1892. Pound was busy establishing his law career between 1890 and 1892, and in 1892 became a partner in his firm. Pound and his friends were without doubt energetic and enterprising, but it would not be surprising if they did not have time to follow European geographical botany. It is likely, therefore, that they were not aware of Oscar Drude's *Handbook of Geographical Botany* of 1890, until after his later work on German phytogeography was sent for review. Even if they were aware of the work, the key members of the Sem. Bot. were not then professional botanists or even graduate students in botany. By 1894 and 1895, in contrast to the 1890–92 period, the Sem. Bot. had become professional. The undergraduate students had advanced to graduate status. The botanical survey was formally under way, with funding for summer research trips paid by the Department of Agriculture and some printing cost to be later covered by the state legislature. Pound and his friends had been thinking about geographical botany from the perspective of applied research problems for years with their teacher.[42]

What did Oscar Drude give to the Nebraska botanists? How did it stimulate Pound and Clements to take a quantitative approach to the formation of the Great Plains? Drude provided them, first, with a set of definitions that were logically related, especially the definitions of "habitat" and "formation." He provided them, second, with a somewhat mechanical theory relating physical and biological factors that create the form of plants and their geographical distribution. He provided them, in other words, with the means of uniting their Bessey training in plant physiology with their applied research concerns in the botanical survey. As if this were not enough, Drude attempted, in the 1890 *Handbuch*, to describe the vegetational boundaries of North America, utilizing his theoretical framework. In their enthusiasm for Drude's theory, they attempted to apply it to Nebraska. They noticed one matter immediately: Drude's demarcation of vegetational boundaries in the Great Plains in the 1890 book, which was based on catalogs of touring botanists and herbaria like their own, was wrong.

Drude divided the North American continent into fourteen regions, which included the following divisions for the prairies: the northern forest-prairie region, the Missouri prairie, the steppe and salt-waste region of the Rocky Mountains, the deciduous forest region along the Mississippi River, and the chaparral region of Texas.[43] From the perspective of Pound's and Clements's intimate knowledge of the prairies of Nebraska and the botanical survey's accumulated records of plants in the state, these geographical regions were grossly inaccurate. Drude had never visited the Great Plains, but relied, instead, as botanists traditionally did, upon lists of species from collectors to define the floral boundaries.[44] Pound and Clements immediately took mathematical exception to Drude's formulation. While Drude had not included eastern Texas and Oklahoma in the province of the Mississippi Forests, Pound did. He and his colleague used a 20 percent wooded criterion. While Drude distinguished between the Canadian and Missouri prairies, Pounds and Clements did not. Drude did not include the Rocky Mountain foothills in the Great Plains; Pound and Clements did. In general, Drude had perceived the North American prairie as smaller and more limited in extent than did Pound and Clements.

How could Drude have made the Missouri prairie smaller than it really was, according to Pound and Clements? For the Nebraska scientists, Drude's mistake was based on the mistake of the collectors from whom he obtained information. Theirs was the mistake of *impressionism*. They characterized the Great Plains in terms of certain species of plants that they viewed while traveling and collecting in it. Plants and flowers that appeared to the eye to dominate the prairie vegetation did not *actually* dominate it. Here is how Pound and Clements stated the error of impressionism:

> In determining the abundance of species, appearances are extremely deceptive. One who has worked over the prairies for many seasons comes to think that he can pick out instantly the most abundant secondary species. Long continued observation in the field stamps a picture on one's mind, and it seems a simple matter to pick out the several species and to classify them in the several grades of abundance with reasonable accuracy. As a matter of fact, this is not possible.[45]

"Actual," "as a matter of fact"—phrases like these say directly that the authors have found a true method, one that reveals a reality behind the falsifying impressionism of the naked-eye observer. What provides access to that reality?—counting. The botanist must count the plants on the prairies, one by one, to know for certain which is most abundant and the grade of abundance into which each falls. Pound and Clements wrote that they made the mistake of impressionism for ten years, before they realized that they had to count plants to know how many there were:

> After more than ten years of active field work on the prairies, it seemed to the writers that the mental pictures acquired was [*sic*] approximately sufficient to make the reference of the commoner secondary species of prairie formations to their proper grades an easy task. When actual looking at the prairies as the season permitted appeared to confirm the picture already formed, this seemed certain. Closer analysis of the floral covering proved that the conclusions formed from looking at the prairie formations and from long field experience, without actual enumeration of individual plants, were largely erroneous.[46]

This long statement contains a revealing mixture of autobiography and psychological testimony. They had worked for ten years, from 1887 to 1897 (assuming that this article is a version of the paper delivered to the Sem. Bot. in the fall of 1897), forming mental pictures. That is, they had based their categories of species abundance directly upon sensations, just as Drude, in the 1896 German phytogeography, said that botanists should. The statement appears to say that they had occasionally worked from herbarium specimens, presumably during the writing of the *Phytogeography*, assigning species to their grades of abundance by relying on their memory's pictures of prairie scenes of these species. In the spring or summer, "as the season permitted," they journeyed to the field, where their memory's mental pictures were confirmed by reviewing the scene.

But these mental pictures, even apparently reconfirmed, turned out not to be accurate. What suspicion should have entered Pound's and Clements's minds that the pictures were not accurate, and that they should have to count individual plants to understand how the prairie composition changed geographically? The first clue that impressionistic pictures were not accurate came from Drude. By relying on other traditional botanists, Drude relied on the same methods as did they, except that he did not actually visit the prairies. And Drude's regional boundaries for the midwestern prairies were wrong to Pound's and Clements's experience. Yet relying on the same methods, shouldn't they have derived the same boundaries? The second clue came in trying to apply Drude's methods to obtain their own, new, regionalization of Nebraska's flora, based on climate and soil. The two scientists explained the problem in the *Phytogeography of Nebraska*.

> Actual field experience has shown that species which appear most prominent in the constitution of the prairies, even to the careful observer, are not necessarily the most abundant. Thus the prominent-flowered blazing stars and prairie clovers make a much greater impression on the eye than other species which are really more abundant.[47]

This misleading impressionism was particularly difficult when the prairie grass formation slowly changed itself into another formation, as in approaching the sand hills in the central part of the state. "Thus, in tracing the shading-out of the prairie-grass formation as we approach the sand-hills, the change in the grade of abundance manifested by the principal secondary species affords a striking character."[48]

The shading-out of the prairie grass formation presented a particularly difficult problem in applying Drude's principle that vegetational formations were distinguished biologically, not simply politically or physiographically. The decrease of the abundance of secondary species was so subtle that it was invisible to the passing naked eye, to the botanist crossing the state by train. It was invisible, yet as real as the slow sweep of the electric clock hand. Or, to consider a problem of even greater difficulty: how was one to draw the boundary on a map between the different "types" of prairie grass formation? How did one define the transitions, from "the prairie grass formation of the prairie region to the buffalo grass and bunch grass formation of the transition area and of the sand hill region."[49]

In two separate publications, Pound and Clements jointly pointed to the problem of delimiting the boundary between the prairie grass formation and the buffalo grass formation as an example of the need for "actual enumeration" of the individual plants. This emphasis reveals the specific historical problem that led to meter-plot. Certainly, Pound had thoroughly botanized the approach to the sand hills and the sand hills themselves. Clements remembered that it was during a minutely detailed survey of the sand hills in the central part of the state in 1892 that Pound and Smith thought up the idea for a formal botanical survey, and Clements added an anecdote of another trip, by train, to the sand hills.[50]

Drude had stated that formations were biologically bounded. In their main, the prairie grass and the buffalo grass formations were recognizably distinct. In traveling between the two, however, it was impossible impressionistically to see a boundary between them, to see the subtle shift in preponderances of the principal secondary species, because the secondary species were so showy that, to the eye, they did not appear substantially to change. The theory in principle predicted a real biological boundary that was not apparent to the eye. Pound and Clements must have realized first that the real biological boundary was invisible, and second that it could only be discovered by counting plants.

Count individual plants of grass on the visually limitless prairies? How could that be done? When would one stop counting? The answer to the problem of counting, and the method by which the two scientists settled on the five-meter square plot was simple. In two separate places, Pound and Clements implied that they simply started counting plants and counted as many as they could in one day. After several attempts, they discovered that on the average, a plot of five-meters square could be handled in one day. In his memoir on Frederic Clements, Roscoe Pound recalled, "In particular, I remember how we worked out together by trial and error in the field the 'quadrat method' described in our paper 'A Method of Determining the Abundance of Secondary Species.' "[51] In affirmation of Pound's memory that they had not simply learned or obtained the meter-plot from Drude, as the historian Rodgers believed, Clements noted that the quadrat method was "devised," surely implying it was worked out, not simply borrowed and applied.[52] Finally, in the *Phytogeography*, Pound and Clements pointedly wrote of the practical reasons for adopting the five-meter square plot, reasons that referred historically to the trial-and-error efforts to develop a standard plot. "The plot used is as large as can be adopted consistently with accuracy in counting."[53]

One meter-plot, of course, would not establish a biological boundary for a floral region. Pound and Clements therefore placed plots in "widely separated stations in the same district," attempting always to situate the plots in "typical situations."[54] While it was not exactly clear, in either their 1898 or 1900 publications that discuss the meter-plot method, what they thought was the appropriate statistical methodology, they were quite aware that "the deficiencies resulting from the small size of the plots are corrected by taking a large number of plots at each station and averaging the results."[55] Meter-plots strung east-west across Nebraska quantitatively defined the transition from one kind of prairie to another. Slowly, across the rolling prairie hills, the dominant individual plants declined in numbers per meter-plot, and were replaced by other dominants. This shift, subliminal to even the most astute observer, was now revealed for the reality it was.

The invention of a quantitative method for ecology was more than the clever application of statistics. The invention of the quadrat, or meter-plot, embodied a profound epistemological shift, in which the scientists ceased to believe in the reality of one phenomenon and began to believe in the reality of another phenomenon. Ecology had "taken leave of its senses," and hitched its intellect to mathematics. It was a shift analogous to the shift in astronomy from the Ptolemaic earth-centered observational astronomy to the Copernican sun-centered astronomy, or the shift from Aristotelian physics based on the phenomenal qualities of motion to the Galilean physics of hidden mathematical laws that lay, as reality, behind the phenomena. And even more pertinently, it was a shift analogous to Darwin's shift of evolutionary theory to a statistical theory of natural selection; it was, in fact, a continuation of that intellectual shift initiated by Darwin. Darwin had discovered the continuum for biology; Pound and Clements applied it to ecology.

Pound's and Clements's conception of the biological boundary between formations was a consequence of the Darwinian revolution in biology. It was not simply borrowed from Drude. Neither scientist directly acknowledged any indebtedness to Charles Darwin in their development of the quadrat or the new approach to ecology, and it is not clear that they were consciously aware of themselves as bringing to ecology the Darwinian approach.[56] Nevertheless, Darwin had accomplished in his evolutionary theory the intellectual shift that Pound and Clements, a generation later, accomplished in plant ecology. The German tradition of plant geography, with its ties back to Humboldt, retained a vestige of pre-Darwinian thinking that lingered even in Drude and other floristic botanists. Darwin had destroyed the sensory typology that had underlain classification in biology; he had shown that species were not "types" of organisms, distinguished qualitatively from other species. Species were categories of individuals that differed from one another along a continuum of variation and whose biological boundary of interbreeding was constantly in flux as environmental conditions changed. The typological definition of species, which Darwin destroyed, was based on the assumption that our senses (vision, primarily) had the ability to perceive biological boundaries or to perceive the end of variations of individuals. Our senses could be trusted; the distinctions we saw were the distinctions present in reality.

In *Origin of Species*, Darwin destroyed not only the typological thinking regarding species, but also the epistemology upon which that thinking was based. From the Darwinian perspective, reality was always too complex to be perceived completely.

> Man can act [in artificial selection] only on external and visible characters: nature cares nothing for appearances, except in so far as they may be useful to any being. She can act [in natural selection] on every internal organ, on every shade of constitutional difference, on the whole machinery of life.[57]

Sensory impressions could not be accurate, could not completely reveal to a person the full range of variations that distinguished one individual organism from another. The "tangled bank" of nature was too complex for impressionism to be accurate. After Darwin, the scientists' or naturalists' methodology could never again be the tourism—casual or arduous—of the traditional survey. The tangled bank could be grasped only slowly, by following one function of an animal through all its consequences. Pound and Clements revealed their epistemological shift, in this Darwinian manner, in their language. They wrote that appearances were "extremely deceptive." Their "long field experience" was "largely erroneous." The "fact" of nature was different from their impression and could only be revealed by "actual" counting of plants. What species "appeared" to be the most abundant were not "necessarily" so. The senses could not be trusted. Only mathematics revealed reality.

Implicit in Darwin's conceptualization of species and natural selection was the notion of the continuum. If life forms were not different categorically, then they could only be different along a continuum of degrees of variation. Natural selection acted to favor one variation and suppress another along this scale. As the biological boundary between species, interbreeding shifted back and forth along the continuum of variation, always moving the structure of the species into closer functional fit with the environment. In these terms, the stable boundaries between species, whether plant or animal, and the stable geographical boundaries between vegetational forms were only *apparently* stable. Our senses were too crude to detect the annual shifts between the prairie grass formation and the buffalo grass formation, as they responded to alterations in rainfall, temperature, and faunal impact. Only when this "silent and insensible" shift had progressed greatly did it appear within the range of our sensory faculties.[58] Typological perception of the natural universe was the direct result of the crudity of our senses and falsified nature, which was always subtler and more complex than we could ever directly perceive.

The Quadrat Method in Full Bloom

Within a decade of its invention, the quadrat method had been developed by the Nebraska scientists into the main method of scientific ecology. The name "meter-plot" was dropped after 1900, in favor of the name "quadrat," which had previously been reserved, in line with Drude's useage, for large cartographic areas the size of U.S. government survey sections (nearly township size). It was placed by Clements within a framework of statistical and causal theory. The *Phytogeography of Nebraska*, by Pound and Clements, published in 1900 (revised second edition), became the landmark study in geographical botany in the United States and the first full representation of the meter-plot method. Five years later, Clements published the first American textbook in ecology, *Research Methods in Ecology* (1905), utilizing a score of mechanical measuring instruments and analytical statistics and graphs. The ten years between 1896 and 1905, constituted the watershed of American plant ecology. The meter-plot and the philosophical foundation of the specialty embodied in it were the bases of the new specialty.

Clements's *Research Methods* was the "account" of the methods he and other Nebraska ecologists developed from 1897 to 1905, that is, from the completion of his doctorate and during most of the years that he taught at Nebraska.[59] In these years, he came to perceive the quadrat as more than a technique for establishing biological boundaries of vegetational formations. The string-bounded, small plot method made possible the intensive examination of the plant habitat. Also during these years, Clements moved away from the floristic study of plants which represented Drude's influence toward the dynamic ecology that marked his mature thought, and especially his paradigm-setting work, *Plant Succession* (1916). The development of Clements's dynamic theory is taken up in the next chapter. Here I examine the full mathematical quality of the foundations of ecology in this golden age of the Nebraska group.

The intellectual basis of Clements's study of the plant habitat was the appreciation for the subtlety of natural variation, the same appreciation that underlay Pound's and Clements's drawing of regional, floristic boundaries. In 1897, Pound and Clements had realized that prairies varied considerably, even though to the eye there was no change, as one moved up the slope of the Great Plains. So in 1905, Clements wrote, "habitats differ in all degrees, and it is impossible to institute comparisons between them without an exact measure of each factor."[60] The *Phytogeography of Nebraska* distinguished between plant geography, which concerned the geographical distribution of species and the abundance of individuals, and ecology, which was the relation of the plant physiology and abundance to the factors of the habitat that determined, in a Darwinian sense, those characteristics. The two studies were theoretically linked; for Clements, the *Phytogeography* established plant geography and the determination of abundance, and *Research Methods* established ecology.

According to Clements, eight major physical factors controlled habitat conditions: water content of the soil, humidity of the air, light intensity, air and soil temperature, precipitation, wind, soil class (e.g., sand, clay), and ground physiography. With the exception of soil class, all factors took interval values and were described on a graph. Even soil class could be described in terms of interval values, if the three components of soil—gravel, sand, and clay—were continuously measured. These physical factors had been the object of broad study in plant geography since Humboldt; and, of course, in all botanical and agricultural observations since the ancients the general importance of these factors for the types of life forms had been noted. Clements's originality was in perceiving these factors in a continuum, rather than in qualitative classes, that is, both in deepening his and Pound's insight of 1897 and in extending to ecology one of the fundamental features of Darwin's revolution in biology.[61]

Clements's experience in 1897 of the breakdown of impressionistic regionalization of the grasslands and of grading abundance of showy flowers prepared him to understand the importance of the continua in physical habitat factors. Any effort to apply the Darwinian program to the plant-habitat relationship had to begin with the assumption that the adaptation of the plants to their environment occurred so subtly that only exact measurement of each factor in the habitat would enable the scientist to relate change to the proper factor. As Clements said, "one cannot trace the adaptations of species to their proper causes unless the quantity of each factor is known. It is of little value to know the general effect of a factor, unless it is known to what degree this effect is exerted."[62] Modern ecology, which Clements wished to establish, was not an outgrowth of nineteenth-century naturalism. Nineteenth-century naturalism was based on the fundamental assumption of the adequacy of sensory impressions for understanding biological nature. Pound and Clements had followed Darwin in rejecting this assumption. For Clements, the ecologist did not study nature, he studied data about nature—data that revealed what he could not see.

Conceptualization of habitat factors in terms of interval values along a continuum immediately led to the use of graph methods for representation and analysis of empirical data. The water content of soil at a particular station over the course of a year was easily graphed with time as the independent variable and water content as the dependent variable. Or—to continue this example—using a line of stations strung east-west across Nebraska as the independent variable, water content could be the dependent variable.[63] The analysis of the relative influences of two or more factors could also be graphed. Thus, temperature, water content, and humidity might be graphed as three different variables. Clements did not deal with the problem of establishing a single scale for all three (e.g., by conversion of their absolute values to percentages), but did state that curves so graphed could be compared in terms of their slopes, if not their values. In other words, the scientist could visually analyze the rates of change; since this would reveal what factors were more constant and which in greater flux, the foundation was laid for relating changes in plant physiology (i.e., adaptation) to the influential habitat factors. So even without absolute values, such analysis would give important information about a habitat. Clements envisioned a single graphic portrait of a habitat, with most all factors (those with interval values) displayed.[64]

Precise measurement of physical factors of stations and of plant species depended upon the quadrat. As he accurately stated, quadrat methods "constitute a satisfactory system, if not, indeed, the only one for exact study of formations."[65] If not, indeed: Clements even charged that descriptive methods in ecology, referring to traditional observational methods, were making "actual progress in the field of ecology more and more difficult."[66] He was aware that the quadrat methodology was tedious and perhaps pushed amateur ecologists out of the field by taking away the pleasure of field research. The method seemed "like drudgery to the mere dabbler" in ecology.[67] Clements did not apparently mourn the exclusion of nature study enthusiasts from professional ecology and he clearly considered the trade-off a good one for the profession. With the quadrat method, ecological knowledge could progress. "In no case has the writer ever listed or mapped a quadrat without discovering some new fact or relation, or clearing up an old question."[68]

Professional criticisms of the quadrat were not easily disregarded. Clements said that one important criticism was that the quadrat method provided only knowledge of a small portion of an immense area, of only five meters square compared to thousands of square miles of prairies, for instance. His response was quasi-statistical. If the vegetative formation were real, and not imaginary, then a quadrat taken anywhere within it would be "in some measure representative."[69] He did not define representativeness statistically as either random or typical; but the *Phytogeography* five years earlier specified that meter-plots taken around different stations should be averaged, and we may suppose that Clements meant average (mean) when he said representative.

As fully developed, the quadrat method was "in its simplest form . . .merely a square area of varying size marked off in a formation."[70] All the plants within the square were counted and their locations plotted on a grid-graph representing the square. In practice, the scientist used meter tapes along the sides for accurate location. A fifth meter tape was run across one side. Individual plants along the fifth meter tape were counted and located by Cartesian coordinates. When all the plants along the tape had been located, the tape was next moved down the quadrat a few centimeters for counting and plotting the next line of plants along it. This procedure continued until the entire quadrat was counted.[71]

Clements described five kinds of quadrats: list quadrat, chart quadrat, permanent quadrat, denuded quadrat, and aquatic quadrat. In the list quadrat, the absolute number of individuals of each species present were listed; in the chart quadrat, they were located by coordinates. The permanent quadrat, as its name implied, was permanently situated in a formation and the inhabitats counted and plotted regularly, so that changes from season to season or year to year would be precisely known. Of course, such data could be correlated to physical factors of the habitat, such as rainfall and temperature.

The permanent and denuded quadrats were keys to the study of successional ecology, that is, of the historical movement of the vegetation. Clearly, the quadrat would indicate the formational changes. The denuded quadrat, however, provided an *experimental* basis for investigating succession. Within the quadrat, the soil surface was denuded of vegetation. Then daily or weekly counts and plots were made of the invasion of the area (technically called a secondary succession) by the plants of the formation. Competition between plants, chronological succession and association, displacement, and alignment with the surrounding formation could be examined in detail.[72] To say that the quadrat was key to understanding succession is, however, to leap over a historical point. In the *Phytogeography of Nebraska* of 1900, the concept of succession was not clearly integrated into the ecological theory. It was the development of the quadrat, along with the mounting literature on ecology, that led Clements to establish ecology on the dynamic foundation of succession.

4
FREDERIC CLEMENTS'S
Theory of Plant Succession

"The Soul of a Sensitive Poet and Dreamer"

As a student, Fred Clements worked with such intensity that sometimes he forgot to eat and sleep, and more than once this deprivation produced hallucinations. Bessey sighed with relief when his young colleague married Edith Schwartz, a language student from Omaha. Edith "humanized" Frederic; as she later testified, "someone had to remind him" about the necessities of life. So she remained closely at his side throughout his career, ministering to the special needs of his health, organizing their household, driving their touring car, typing his dictated descriptions of landscape, coauthoring scientific papers, drawing the illustrations of flora; she was a scientist in her own right (with a Ph.D. in botany in 1904). For Edith, of course, who could see the personal side of a scientist who was always somewhat aloof from his professional colleagues, Frederic was more than a man of science, he was a poet and dreamer.[1]

If Frederic Clements did not appear as poet and dreamer to other scientists, he did, nonetheless, appear as philosopher. Without significant historical anticipation, Clements became the first philosopher of ecology. His major principles—fantasies, if not dreams, to his critics—shaped the conceptual development of American plant ecology, helped formulate the basis of American animal ecology, and framed interpretive controversies for generations of ecologists. His name appeared as author or coauthor on landmarks of the profession: his and Pound's *Phytogeography of Nebraska* (1898, 1900), *Research Methods in Ecology* (1905), *Plant Succession* (1916), *Plant Ecology* with John Weaver (1929, 1938), *Bio-Ecology*, with Victor Shelford (1939). His student, Raymond J. Pool, memorialized him without exaggeration: "He forged an amazing career as one of the world's outstanding figures in the general field of biology and in the rapidly evolving specialty of plant ecology. He was an unusually bold, brilliant pioneer as he set out to help to formulate the fabric and to mould the trend of America thought in teaching and research in plant ecology."[2]

Appropriately for an intellectual pioneer, Clements lived most of his life on the frontiers of America. He was born in Lincoln, Nebraska, in 1874, where his father, a photographer, maintained a studio. It was the same new and raw Lincoln that greeted C. E. Bessey upon his arrival. The city was not ten years old, the surrounding prairie was unbroken. The Indian Wars with the Sioux and General George Custer's last battle occurred two years later less than five hundred miles away. Clements enjoyed the outdoor life and outdoor sports. He played football in high school and at the university, and was a cadet officer at the university when Pershing was commandant. He so enjoyed this activity that he frequently sported military-styled clothes—jodhpurs, laced field boots, campaign hat—when he was in the field.[3]

His intellectual and cultural interests ranged widely at the University of Nebraska, but he focused on botany under the influence of Bessey's teaching and the inspiration of the many-faceted Roscoe Pound. In an unpublished memoir, Clements's admiration for Pound shines brightly; in "emulation of his example," Clements began his collecting activities.[4] To be Pound's friend was to be caught in Pound's whirlwind. Clements entered the Sem. Bot. in 1892, participated in the Botanical Survey of Nebraska initiated by Pound and Smith that year, and collaborated in research and co-authorships, culminating in their landmark, joint doctoral thesis in 1897. Pound stimulated Clements in key intellectual directions. As we have already seen, Pound's reading of Drude's *Deutschlands Pflanzengeographie* led the young men toward ecology. Pound and Clements jointly moved to the quantitative quadrat method. And Pound read and discussed with Clements the major philosophical issues relevant to biology—Linnaeus, Lamarck, Darwin, Spencer. Pound, the Latin punster of Nebraska, apparently awoke in Clements the thirst for languages and concern for linguistic exactitude (an interest reinforced by Edith Schwartz, who held a job as a teaching assistant in German in 1898, the year they were married). Pound, teaching as well as practicing law in the late 1890s and early 1900s, identified himself with the "sociological" school of the law in the years sociology would apparently influence Clements's conception of vegetation. The paucity of manuscript evidence and testimony from this period of the men's friendship does not allow us to state who gave what ideas to whom; with such a close intellectual partnership, perhaps this is not even a legitimate question. But Pound clearly stands, along with Bessey, as an admired, emulated, towering influence on the young Clements.[5]

After receiving his doctorate in 1898, Clements began teaching in Bessey's botany department, and immediately revealed the Besseyan penchant for laboratory and field studies, as distinct from textbook instruction. He wrote a laboratory manual (coauthored with Cutter) for high schools which influenced Raymond Pool, later department head of the Nebraska botany department.[6] In the summer, 1898, he and his bride vacationed in the Colorado Rocky Mountains, where he enthusiastically took up forest ecology. This new interest led to his conception of the vegetative formation as an organism, which he developed in two books, *The Development and Structure of Vegetation* (1904), and *Research Methods in Ecology* (1905).[7] Publication of these books brought Clements international fame, and he was quickly offered a position elsewhere, the chairmanship of the University of Minnesota Department of Botany, to succeed Edwin McMillan, another Bessey student. Bessey was unable to obtain a sufficient raise for his rising star, and Clements left. His departure coincided with that of Pound, who joined the faculty of the Northwestern University School of Law.[8]

While administering botany at the University of Minnesota, Clements built toward "his big book," *Plant Succession*. The work grew out of Clements's earlier concern with the ecology and development of forests, and represented a formal systematization of the philosophy of the organism and its development initially presented in 1904.[9] Evidently, his efforts to deal with the forest as an organism led him to attempt to validate the organistic philosophy on other formations, principally the familiar prairie. As we shall see by the end of this chapter, Clements's organistic philosophy of the vegetative formation was self-consciously conceived as an alternative to another model of vegetative development being contemporaneously sketched at the University of Chicago by H. C. Cowles.

The fifteen-pound manuscript of the book was read by D. T. McDougal of the Desert Laboratory of the Carnegie Institution and at his suggestion was submitted to the Carnegie Institution for publication.[10] This shortly brought an offer to join the institution as a research associate, a position that Clements accepted and held, though not always without its troubles, until his retirement in 1942.

Although this research position meant withdrawal from teaching, Clements maintained his association with the University of Nebraska. The Alpine Laboratory, established in 1900 in the mountains above Manitou Springs, Colorado, had always been used during the summers for special students. In 1917, the Carnegie Institution took over the funding of the laboratory, and Clements brought Nebraska graduate students to the augmented facility.[11] In 1925, the Clementses moved from Tucson, Arizona, where he had worked at the Carnegie Institution's Desert Laboratory, to Santa Barbara, California, where he established a coastal ecology laboratory and conducted the remainder of his career.

The Philosophy of Vegetation

Plant Succession synthesized earlier contributions to understanding vegetational change and set them in a logical system with sufficient generality to make his theory the microparadigm for plant ecology. Clements's ally and the founder of British plant ecology, A. G. Tansley of Cambridge University, immediately recognized these qualities in the work, hailing it as the basic theory for the examination of developmental change in vegetation.[12] The synthesis was based, however, on a philosophy of vegetation that was immensely controversial. Independent of Clements's theory, plant succession meant that all vegetation was constantly in change: plant communities grew, matured, declined, and were replaced by other plant communities. Moreover, these changes always occurred in a structured sequence. Associated animal communities were transformed by the support vegetation. In Clements's theory, all plants were organized into large units of vegetation called formations, such as forests, which were themselves organisms. In addition, the change that these formational organisms underwent was irreversibly directional. Formations developed toward the climatic climax, which was the matrix of plant and animal life most suited to a climatic region, such as the grassland-buffalo biome in the presettlement, North American Missouri plains. This dynamic theory of directional change in the living world was accepted and utilized in the United States and Great Britain, where it had early adherents, but was largely rejected by ecologists on the European continent, who had a more static view of their native habitats.

Clements's vision was fundamentally ambiguous, expressing both an idealistic interpretation of growth and its contrary, a mechanical model of change. In *The Development and Structure of Vegetation, Research Methods in Ecology*, and *Plant Succession*, Clements remained stalwartly committed to the conception of the vegetational formation as an organism, growing through seral stages to maturity.[13] Occasionally he explained his conception in analogy to embryological development, repeating the formula of romantic animal embryology that "ontogeny includes both the ontogeny and the phylogeny of climax formation."[14] The embryo went through a series of stages, according to some early nineteenth-century biologists, repeating the stages of development of the species of which the embryo was a member. Similarly, the succession of plant groups went through a staged sequence in becoming the final adult form or terminal formation. So, for example, on a denuded patch of contemporary prairie soil, the prairie would not heal itself by immediately filling in with mature prairie; rather, the patch would progress through different stages of grass and shrub mix until the composition of the mature prairie was reached. Clements's conception of change was typological, therefore, by providing a schematic for analyzing growth in terms of distinct categories or types of being, with succession as the change of one type into another.

Clements's theory also contained a population concept, which was modern, or Darwinian, in its thrust, and represented the influence of Bessey. This population concept was utilized, as we saw in the previous chapter, from his and Pound's quantitative understanding of vegetational change. From this point of view, the difference between one stage and another in succession was often not expressed qualitatively, that is, not expressed visibly to the human senses. Rather, the transition between stages was a subtle shift in numbers of individuals of different species in the total population of plants. Detection of this shift required the quadrat method. As late as 1916, Clements alluded back to the experience of the 1890s in analyzing plains habitat:

> Even where the final community seems most homogenous and its factors uniform, quantitative study by quadrat and instrument reveals a swing of population and a variation in the controlling factors.

Invisible as these are to the ordinary observer, they are often very considerable, and in all cases are essentially materials for the study of succession.[15]

This population concept relied on the Darwinian vision of evolution, in which a population of reproducing individuals were acted upon by habitat and biotic factors in such a way that the favored had higher reproductive and survival rates than the less favored, and one subpopulation of individuals slowly replaced another subpopulation in the larger composition of which they were parts. Clements wrote that succession's "most striking feature lies in the movement of populations, the waves of invasion, which rise and fall through the habitat from initiation to climax."[16]

The two conceptions of formation—organism and population—implied quite contradictory conceptions of development. The organismic model, on the one hand, implied that development was caused by a major external cause (not to say goal). The plant formation was pulled toward its climax by the climate, which it did not of course initially create. The habitat was, therefore, of less causal importance than the climate. The population model, on the other hand, implied strong habitat influence on the development of vegetation. The climate was of secondary importance, and it was conceivable that habitats should produce a variety of terminal formations under the same climatic influences (indeed, British and Continental ecologists were certain that this was the case). The population model was conceptually mechanical, just as the organismic model was, contrarily, embryological. Drawing upon the classical concept of mechanical equilibrium, Clements described the habitat and plant population as interacting, to paraphrase him, "alternating as cause and effect until a state of equilibrium is reached."[17]

Clements thought that all the different viewpoints on vegetative development were synthesized in the organismic model, mechanical metaphor not excluded. "All of these viewpoints are summed up in that which regards succession as the growth or development and the reproduction of a complex organism."[18] Adoption of the organismic model was not a matter of heuristic convenience for Clements, as it would become for A. G. Tansley, who in 1931 referred to the "quasi-organism."[19] For Clements, the formation was ontologically as real an organism as the individual plant or animal. Hence: "[Succession] is the basic organic process of vegetation, which results in the adult or final form of this complex organism."[20] And: "All the stages which precede the climax are stages of growth."[21] As he stated in the second sentence of *Plant Succession*, "As an organism the formation arises, grows, matures, and dies."[22]

North America has been ideally suited to perception of major formations of vegetation. The different vegetative covers are strikingly differentiated and correspond to large climatic regions. In reviewing *Plant Succession*, Tansley noted this feature of the continent, implying that the appreciation of the concept of the formation as a climatic community by Western European botanists was stifled by the uniformity of the European climate.[23] Clements distinguished between three major American formations: the deciduous forest climax of eastern North America, the prairies-plains grassland climax of the midcontinent, and the Cordilleran climaxes or coniferous forest climaxes of the Rocky Mountains. Other climax formations, such as the desert, were not major. The primary climaxes were the long-term, stable, final formations for the continent in existence since the recession of the last glacier. Drought, wet years, and even cultivation could do no more than deflect the growth of vegetation toward these formation types, as long as the climate remained basically stable.

Development of vegetation toward its terminal climatic formations was always progressive. It could not permanently regress or stall short of the final form. "Succession is inherently and inevitably progressive. As a developmental process, it proceeds as certainly from bare area to climax as does the individual from seed to mature plant."[24] Although artificial regression, as in clearing a forest, may appear to be a real regression, Clements maintained that "there is nowhere evidence that this is the case."[25] Phenomena like meadow clearings in forests were often interpreted by other botanists as "regressions" away from the climax, but for Clements they were only "subclimaxes," artificially maintained short of the final formation. Succession could no more regress to an earlier stage than a butterfly could reverse itself into the pupa.[26]

Where and how did Clements obtain the peculiar mix of views in his philosophy of vegetation? The origins of the mechanistic-Darwinian metaphors are obvious enough: The *Origin of Species* transmitted through the great nominalistic Darwinian, C. E. Bessey. But the sources of the organistic model are not equally obvious. I believe we can discern two distinct influences on Clements's concept of the formational organism. The first is the sociological tradition of Herbert Spencer and Lester Frank Ward, which avowedly conceived of human society as a superorganism, or organism-in-itself. The second source is the idealistic tradition in plant geography, represented by Alexander von Humboldt, August Grisebach, and Oscar Drude, which conceived of the vegetative formation as a unit ontologically distinct from the individual plants comprising it. Since the idealistic tradition in plant geography represents a professional botanical influence culminating in Pound's and Clements's *Phytogeography of Nebraska*, under the inspiration of Drude, I deal with it in a separate section following the present discussion. Here I examine the sociological tradition and the evidence for its influence.

In *Research Methods in Ecology* (1905), Clements informs the reader that his concept of vegetation as a complex organism arose out of his research on the Colorado forests.[27] He initiated this research no earlier than the summer of 1898, when he and Edith joined a party of vacationing teachers at a rented cabin below Pike's Peak.[28] Frederic quickly envisioned a research laboratory in the mountains. For the next thirty-four years, he and Edith usually returned each summer to their forest station above Manitou Springs. His interests in forest ecology bloomed prodigiously; at the August 1901 joint meeting in Denver of the American Association for the Advancement of Science and the Botanical Society of America, he presented five papers, one on plant formations in the Colorado Mountains and four on various aspects of ecology.[29] He developed the concept of the vegetational organism, therefore, between 1898 when he first visited the Rocky Mountain forests and 1904, when he elaborated the philosophy in the *Development and Structure of Vegetation*.

In thinking through the problems of Colorado forests, considered as vegetative formations, Clements after 1898 naturally drew upon the tradition of plant geography represented by Drude and so influential in the *Phytogeography*, published in 1898 and revised in 1900. What evidence is there that he now picked up a Spencerian notion of the societal organism and used it also to help think about forests? Why should he have been thinking about sociology at all in the years between 1898 and 1904? Clements's acquiescence in another botanist's narrative of the history of the concept of the complex organism is our best indicator of the likely influence of the sociologists. In 1935, the South African ecologist, John Phillips, published an article on the complex organism in which he identified the sociological tradition of Comte, Spencer, and Ward as the major course of the concept. Without stating directly that Clements borrowed the concept from the sociologists, Phillips described Clements as the first thinker to apply the concept to plant communities. This discussion clearly implicated Clements as legatee of the superorganistic tradition in sociology. Clements and Phillips had actively corresponded since Phillips's student days in 1920 and the two scientists had met on Phillips's visits to the United States in 1930 and 1934. Phillips would not have implied a sociological influence had there been none.[30]

Given the great popularity of Herbert Spencer's philosophy and Lester Frank Ward's sociology in the fifteen years before Clements's vacation in Colorado, we could plausibly argue for their influence in almost any year since Clements entered college. Nevertheless, there are reasons to see the years after 1898, and especially after 1901, as bringing a special confluence of interest in sociology at Nebraska. Both Spencer's and Ward's works received renewed attention at the end of the century. Ward's *Dynamic Sociology* was republished in 1897, eliciting a flurry of magazine reviews. It is inconceivable that Clements, reading widely in the company of Nebraska teachers of different fields, would not have been aware of Ward's work. Indeed, the very title, "Dynamic Sociology," bore suspicious resemblance to "dynamic ecology," the label Clements affixed to his own vegetative philosophy. Spencer's voluminous writings were continuously in revision and reprinting; the *Principles of Biology* was republished in 1898 and carried the efforts of younger biologists to whom Spencer had entrusted the task of verifying its biological details. Roscoe Pound remembered reading *Principles of Biology* together with Clements, despite Bessey's warning that they would get nothing out of it (and Pound recalls they didn't). The *Principles of Sociology* (vol. 1, 1876, 1877, 1885) was reprinted in 1904.[31]

In 1901, the sociological connection became tighter. E. A. Ross was recruited to Nebraska from Stanford University. Ross was not only a sociologist in the Ward tradition, but he was a personal friend of Ward. Years later, Ross wrote to Clements to remind him of one especially vigorous discussion of sociology and ecology in 1903 or 1904. And the omnipresent Pound! Pound became an assistant professor of law at Nebraska in 1898, rising to become dean of the law school in 1903. In 1904, Pound published an article identifying himself with the "sociological school of Jurists." How can we doubt, with his closest friend plunging into the sociological literature, that Clements conversed about and read frequently in sociology?[32]

The Spencerian sociological tradition elaborated several principles that appeared in Clements's ecology. Spencer particularly legitimated the conception of human society as an organism. In essays of 1860 and 1873 and in 1880 in his *Principles of Sociology*, Spencer asserted "a society is an organism" (in fact, this is the title of a chapter). Just as the individual organism grows, so the superorganism (Spencer) or complex organism (Clements) grows. To quote Spencer: "A society as a whole, considered apart from its living units, presents phenomena of growth, structure, and function, like those of growth, structure, and function in an individual body."[33] Lester Ward, too, utilized the organistic analogy. *Dynamic Sociology* (1897) refers repeatedly to "society as an organism." If, for instance, society is to progress, men must manipulate the laws that apply to it, just as they have manipulated the laws of breeding to improve stock. "Society is simply a compound organism whose acts exhibit the resultant of all the individual forces which its members exert. These acts, whether individual or collective, obey fixed laws. Objectively viewed, society is a natural object, presenting a variety of complicated movements produced by a particular class of natural forces."[34]

Clements's theory of the relationship between the vegetative formation—conceived as organism—and the climate rested upon a fundamental belief in neolamarckism. Each climate had only one kind of formation ideally suited to it, and the climate pulled the formation through the various stages of its development until that ideal type was reached. This theory of climatic causality downgraded to secondary status the Darwinian mechanism of natural selection (which is more related to habitat than to climate). Clements's neolamarckism remained throughout his life a major philosophical assumption increasingly isolating him from the mainstream of biological theory, particularly in the 1920s when population genetics was established, based on the Darwinian mechanism. As was his philosophy of the vegetative organism, Clements's neolamarckism was obtained from late nineteenth-century sociology and anthropology.

Lamarckism is the view, originated in a general philosophical form by Jean Baptiste de Lamarck in the early nineteenth century, that the more complex species of plants and animals evolved by transformation of less complex species, as those less complex species attempted (by their own effort, in the case of animals) to adapt themselves to new environments or to changes in their own environments. Alterations in bodily form acquired by an organism as it changed in response to changed conditions were transmitted to successive generations. Darwin's theory of evolution by means of natural selection, although indebted to Lamarck's views, was fundamentally different. Rather than the plant or animal adapting to a changing environment, the changing environment, in Darwin's theory, accumulated variations in offspring favorable to survival in the new conditions. Adaptation appeared in Darwin as the result of changing biological form, not, as in Lamarck, the cause of change in form.

Neolamarckism emerged at the end of the century as an evolutionary doctrine that opposed the idea of the power of natural selection. In the opinion of outstanding neolarmarckians, such as Edwin Drinker Cope, ethnologist and founder of *American Naturalist* (for which Bessey was editor of the botany section), "natural selection" was an empty concept. It did not explain the origin of variations. It made evolution too slow. It understated the obvious power of environment to force organisms to change their form and behavior. For the neolamarckians, the relationship between environment and organism was direct; the organism responded, with changes in its chemical and mechanical structure, to forces in the environment, such as temperature, light, and moisture. The power of environment caused rapid transformation of species during periods of rapid environmental change.[35]

At the end of the century, the neolamarckian position became involved with American anthropology and sociology. The neolamarckian philosophy, in the hands of liberals like John Wesley Powell and Lester Frank Ward, implied the power of environment to change human nature. Environment was thereby more important than heredity in determining the biological constitution of individual human beings and the constitution of society. Needless to say, this position promoted the philosophy, especially in the works of Ward, that governmental intervention to alter the environment could promote the progress of society.[36]

Clements formed his views on evolution in the 1890s, when the neolamarckian movement was at its peak. Though Bessey was a trenchant, nominalistic Darwinian, who emphasized the power of natural selection, his scientifically based pragmatism was similar to Ward's. Clements never abandoned his neolamarckism, even when at the end of his career experimental disproof of neolamarckism was pouring in. Although he did not publish a big book on his evolutionary views, he published hints of them. Thus in *Research Methods* (1905), he wrote that new species could originate through adaptation to a changing environment, by the plant's response to environmental stimuli. "New forms resulting from adaptation are like those produced from mutation, in that they appear suddenly as a rule and without the agency of selection. They are different, inasmuch as their cause may be found at once in the habitat."[37] Clements outlined a program of research into "experimental evolution," involving transfer of plants from one environment to another, that he conducted for the rest of his career, amidst his many other interests. In the 1920s, this neolamarckism came under widespread attack; yet Clements privately never withdrew his beliefs in the least. In 1925, in justifying his budget to the Carnegie Institution, he wrote optimistically to its chief administrator, Dr. C. Merriam: "With regard to evolution in the narrower sense, my results this spring indicate that I have found the functional mechanism by which the soma modifies the germplasm and thus leads to the fixation of characters arising in response to environmental stimuli. If this prove true, it is the most fertile lead since Darwin's original thesis." At the same time, he was anxious to deny that population genetics, with its study of the fruit fly—a "house of cards"—could discover anything fundamental about evolution. As late as 1942, he maintained that "we have produced a large number of new forms, some of which would pass for 'new species' in the hands of the systematists, by the direct action of environment."[38]

The Idealistic Tradition in Plant Geography

The first intellectual influence pushing Clements onto the road to ecology was, of course, the plant geography of Oscar Drude, which provided Pound and him with a new perspective on the prairies. As Clements wrote, years later, to Drude, "You will recall that I obtained my first clear view of our field through your 'Deutschlands Pflanzengeographie' and have always remained under obligation to you for this and your other stimulating works."[39] Drude's eruption on the Nebraska scene was more than simply the *Deutschlands Pflanzengeographie*. Drude represented, as we saw in earlier chapters, a professional tradition of which the Nebraskans had little previous inkling. Drude's plant geography culminated nearly a century of German floristics, with a definite conception and mode of treatment of vegetation. However attenuated it may have been in Drude himself, Drude transmitted German idealism to Nebraska in his understanding of the vegetative formation and its succession. Drude and the German idealistic plant geographers understood the formation to be a unit of vegetation ontologically distinct from the plant, to be explained in terms of its own laws. Clements's acceptance of this concept in the years between 1896 and 1898 prepared him for the Spencerian conception of the formation as an organism he obtained in the following years.[40]

Idealism was a major philosophical theme in Western culture in the nineteenth century. For the first part of the century, science, as well as secular philosophy, had a strong Kantian element alongside the Newtonian mechanism. The Kantian philosophy metamorphosed itself into many philosophical lepidoptera: Fichte's and Hegel's idealism; romanticism or *naturphilosophie* in Goethe, Faraday, Schleiden, and Agassiz; and literary romanticism in the romantic poets, especially in Coleridge and Emerson. Academic philosophy in England and the United States, represented primarily by ministers and theologians, loyally hewed to Lockean empiricism and Scottish common-sense realism. At mid-century, good English translations of Kant's *Critiques* became available and German idealism rapidly gained popularity as a technical (i.e., ontological) philosophy among academic philosophers. Josiah Royce, William Torrey Harris, and the young John Dewey in the United States, and F. H. Bradley and the young Bertrand Russell in England converted to the German metaphysics. At the University of Nebraska, technical idealism was represented by E. L. Hinman, who came to Lincoln in 1896 to establish the philosophy department. He received his doctorate from Cornell University in 1906, with a thesis on idealism in nineteenth-century physics. Herbert Schneider, in his classic narrative of American philosophy, believed that Hegelian idealism professionalized American philosophy. Based partially on the assumption that something true can be known about the world by pure thought, idealism gave philosophers a job: to build systems. Consequently, idealists frequently dominated university departments of philosophy at the end of the century. Opponents of idealism sniped away from other departments. Among pragmatists, C. S. Peirce worked for the U.S. government, William James was a psychologist, and John Dewey, who abandoned idealism for pragmatism, was in education. Realists entrenched themselves in literature.[41]

American middle class culture was permeated with idealism, from the 1830s, when Emerson formulated his views, to the 1890s, when transcendentalism was nostalgically repeated by the muted trumpet of the genteel tradition. As we have already seen, this genteel tradition received powerful support from mid-nineteenth-century science, represented especially by the natural history tradition of Louis Agassiz at Harvard. Middle class individuals who were not attracted to technical German idealism absorbed nature philosophy, as a variant of idealism, in Alexander von Humboldt's *Cosmos*. Humboldt's *Cosmos* was immediately translated into English and went through many editions by the end of the century.

In nearly all aspects of British and American intellectual culture, therefore, idealism in some form was a powerful philosophical view. Darwinism—or what we may call realistic naturalism—struggled long to displace idealism among the scientists and no doubt some form of diluted idealism, perhaps simply as secular theism, persisted privately in many scientists after naturalism, or materialism, or functionalism, or realism—by whatever name the opponents of idealism went—had become the new orthodoxy.

To understand more clearly the idealistic language and concepts in Clementsian philosophy, I wish to examine the idealistic tradition in botany. I distinguish idealistic and mechanistic traditions in botany, though I do not want this distinction interpreted in broad and absolute terms. The distinction is based on the historical origins of *naturphilosophie* in botany and on the way the two traditions treated the concept of the "vegetative" formation. The idealistic tradition always gave an ontological status of reality to the formation as a unit, separate from status of the individual plant, while the mechanistic tradition denied this reality. Also, by labeling one tradition ideal-istic, I do not mean that all the participants accepted all the ambiguous implications of idealism or romanticism or the associated technical arguments of the idealistic ontology, as in, for example, Hegel. Rather, after the first generation, the idealistic tradition became a manner of handling certain concepts and of nuances in the language by which theoretical concepts in botany were expressed.

I identify the idealistic tradition in botany with Goethe in the first generation, followed by August Grisebach and Oscar Drude. Alexander von Humboldt, whom Grisebach followed, stood in interesting relation to this tradition. By reason of all his qualities—synthesizer of many scientific fields, author of popular and technical writings, long-lived friend of diverse scientists, participant in social circles of contrary philosophical inclinations—and not least his tendency to write for his audience, Humboldt during his lifetime expressed quite different philosophical perspectives on nature. He was not a technical philosopher and did not always hew his opinions to a careful consistency. Thus, in his youth, he could be sufficiently excited by the chemical revolution of Lavoisier and stimulated by the discoveries of Galvani to flirt briefly with materialism, only to rebound quickly to a vitalistic view of life phenomena.

Two intellectual tendencies obtained in Humboldt's early career remained with him always. One was a solicitous regard for the primacy in science of the empirical fact experimentally verified. The other was the not-entirely-compatible enthusiasm for a universal science, a natural history in which all the empirical facts of the world would be connected by basic explanatory principles. Both of these mental allegiances influenced Humboldt's conception of plant geography. His experimental empiricism anchored itself in a personality chary of criticism about fanciful hypotheses and was reinforced by love of French science and scientists. During his long and happy sojourn in Paris, following his return from South America, he established friendships with Joseph Louis Gay-Lussac and D. F. Jean Arago and participated in the Society of Arcueil around Claude Louis Berthollet. The French scientists did not welcome the *naturphilosophie* brand of subjective speculation, then popular in Germany, maintaining rather empiricism and mathematical positivism. In their company, Humboldt searched doggedly for mathematical laws, particularly in magnetic phenomena, and developed mathematical concepts, such as the isotherm, in plant geography.

When his brother, Wilhelm von Humboldt, moved to Jena in

1794, Alexander came to know the literary and philosophical friends there and at Weimar and entered what his biographer Hanno Beck has called, "the post-Kantian circle." Goethe particularly influenced the young scientist. Alexander testified in a letter to a friend:

> I have been constrained to admit, while ranging the forests of the Amazon, or scaling the heights of the Andes, that there is but One Spirit animating the whole of Nature from pole to pole—but One Life infused into stones, plants, and animals, and even into man himself. In all my wanderings I was impressed with the conviction of the powerful influence that had been exerted upon me by the society I enjoyed at Jena, of how, through association with Goethe, my views of nature had been elevated, and I had, as it were, become endowed with new perceptive faculties.

Goethe's conception of the Ideal Type of plant species and his belief that the plant cannot be understood except through its entire life history, which were strongly influenced by Kant's views on the concept of the organism and teleology, influenced in turn Humboldt. In *Géographie des plantes*, the German edition of which was dedicated to Goethe, Humboldt endeavored to perceive the history of the vegetational region, producing particular plant forms through climatic and mineralogical conditions. He conceived the region, holistically, as having a unity within boundaries measured by the barometers and thermometers and other instruments he laboriously carted over South America.

The concept of the climatic region, which produced unique vegetation and plant forms and has special consequences for human society in it, belongs to an ancient tradition of environmental observation and speculation. These ideas had been recently developed by the great eighteenth-century naturalists, Linnaeus and Buffon, for instance. Humboldt responded to his reading of these traditions, of course, as well as to exposure to Goethe and Kant. He was, additionally, powerfully influenced by friendships. His friend, Karl Willdenow, conceived of plant geography in terms of life history of vegetation, a philosophy absorbed by the younger scientist. Another compatriot, George Förster, was virtually a model in intellect and style for Humboldt. The founder of plant geography cannot be characterized simply as belonging to just one intellectual tradition. He shared diverse currents that swirled through his era of intellectual and political revolution.[42]

In spite of the range of influences on Humboldt and the range of his own opinions, it remains possible to perceive the importance of Kant's views of biological phenomena for him and other naturalists of the post-Kantian circle. In the *Critique of Judgment* (1793), Kant explained that biological nature may legitimately be described as having *ends*. Teleological explanation was legitimate (and inevitable) in the scientific analysis of the behavior of organisms. This teleology was taken over, in modified form, by Goethe and Humboldt. For Kant, organisms were the only beings in nature to whom the concept of teleology was applicable. An organism was "an organized natural product . . . in which every part is reciprocally both end and means."[43] This meant a scientist would analyze an organism by inquiring what end each organ performed for other organs, or when looking at the interactions of many organisms, ask what end (goal) was served by processes like growth and destruction. Scientists could never *prove* that organisms were formed by more than efficient, nonteleological causality, but nevertheless they could not analyze organisms without using a teleological principle to guide or regulate their thinking. "The principle or reason is one which it is competent for reason to use as a merely subjective principle, that is as a maxim: everything in the world is good for something or other; nothing in it is in vain." And: "this is a principle to be applied not by the determinant, but only by the reflective, judgment, that it is regulative and not constitutive, and that all that we obtain from it is a clue to guide us in the study of natural things."[44]

From these Kantian influences, the young Humboldt was brought to Kant's perspective of a harmoniously functioning universe that developed historically (that is, "unfolded" itself). This perspective was particularly appropriate for geological and geographical studies. Hanno Beck, Humboldt's outstanding biographer, wrote of Humboldt's youthful "Florae Fribergensis specimen" (1793): "Wie Kant, un geistgeschichtlich im Zusammerenhang mit ihn, entwickelte Humboldt eine Gliederung der Geographie, die er bis an seim Lebensende beibehielt. Die erscheinungen Können nach drie Gesichtspunkten behandelt werden: dinglich-systematisch, historisch und chronologisch."[45]

Humboldt utilized the Kantian philosophies to which he had been exposed in the 1780s and 1790s in his plant geography and in his developmental theory of the universe in *Cosmos*. He perceived the vegetation of the earth as regionalized according to the climates of the planet and differentiated according to physical factors such as temperature and rainfall. Every climatic region had an appropriate vegetational type, for example, tropical rainforest, subtropical grassland.[46]

The theory of the regionalization of vegetation was tied directly to a developmental view of the earth. Humboldt accepted the hypothesis of cosmic and planetary development, with contemporary geological structures and the organization of life forms historically developed out of earlier structures. In his popular lectures, translated as "Aspects of Nature," he described how the earth's surface was divided into regions with different "physiognomy," or faces. Appearances were generated by the unique features of the region or habitat factors. The physiognomies of the Italian Alps, the British Isles, and the Brazilian tropical forests were different and necessary products of their climatic regions. Uniqueness meant being ideally suited for a region in a way that no other kind of vegetation was. Uniqueness did not mean being totally different from all other regions, or of being one-of-a-kind. Humboldt drew analogy to the types of animals—the "determinate physiognomy . . . in distinct organic beings"—or species.[47] Vegetative formations were unique because they were what they had to be and each could be no other; they were not the result of random chance. In developmental terms, Humboldt's view meant that the biological history of a region was similarly directed toward fulfillment of the ideal forms for that type of region.

Humboldt's typological history of regions had its closest analogy to the botany of his friend, Goethe. For Goethe, growth of the plant was a metamorphosis of the primitive seed-leaf. All the structures of the plant after germination generated out of the cotyledon and were transformations of that first leaf. Each plant grew toward the ideal adult form of its species. Embryological development and later growth were progressive and inevitable. In *Cosmos*, Humboldt referred to the "marvel" of "regular succession" and "an internal and progressive development" of organic forms, both in paleontological and ontological growth.[48] Goethe, he said, had treated this problem with "more than common sagacity."[49]

In a discussion of the production of the typical physiognomy of a region, Humboldt described the phenomenon of "succession" with much the same meaning that Clements would give to it two generations later. The succession of organic forms was intimately related to Humboldt's Kantian view of the universe as historically developing itself in its ideal forms. This long passage reveals his conception of "succession" clearly:

> No sooner is the rock of the newly raised islands in direct contact with the atmosphere, than there is formed on its surface, in our northern countries, a soft silky net-work, appearing to the naked eye as coloured spots and patches. Some of these patches are bordered by single or double raised lines running round their margins; other patches are crossed by similar lines traversing them in various directions. Gradually the light colour of the patches becomes darker, the bright yellow which was visible at a distance changes to brown, and the bluish gray of the Leprarias becomes a dusty black. The edges of neighboring patches approach and run into each other; and on the dark ground thus formed there appear other lichens, of a circular shape and dazzling whiteness. Thus an organic film or covering establishes itself by successive layers; and as mankind, in forming settled communities, pass through different stages of civilization, so is the gradual propagation and extension of plants connected with determinate physical laws. Lichens form the first covering of the naked rock, where afterwards lofty forest trees rear their airy summits. The successive growth of mosses, grasses, herbaceous plants, and shrubs or bushes, occupies the intervening period of long but undetermined duration. The part which lichens and mosses perform in the northern countries is effected within the tropics by Portulacas, Gomphrenas, and other low and succulent shore plants. The history of the vegetable covering of our planet, and its gradual propagation over the desert crust of the earth, has its epochs, as well as that of the migrations of the animal world.[50]

The regional vegetation, beginning from nude soil or rock, passed through a series of stages in which communities of plants succeeded one another. The succession proceeded to the more complex plant types until a vegetative community was formed comprising the highest plant form that the climatic region could support. This is not the theory of evolution, and a modern-day analysis of the botanical meaning of the succession would raise questions Humboldt's popular discussion could not answer. A modern scientist might wish to know, for instance, whether mosses preceded trees on the rocky isle because they were simply there first or because a moss community was biologically necessary preparation for the later communities? Despite the obvious inability of the passage to sustain a modern analysis, the general conceptions of biological change as a process of staged development of communities and guidance of that development by a climatic cause, leading to the terminal plant community ideally suited for the region, would meet large agreement by Clements.

The term "formation," referring to the integrated plant community, was introduced to botany by the follower of Humboldt, August Grisebach (1814–79). First using the term in 1838, Grisebach added recent developments in historical geology to Humboldt's concept of regional physiognomy. While later botanists, including Clements, tended to dissociate Grisebach's concept of formation from Humboldt's concept of physiognomy, Grisebach explicitly stated the Humboldtian concept was the origin of the "formation." In a retrospective essay (1872) on Humboldt's contributions to plant geography, Grisebach demonstrated how the concept of physiognomy linked climatic influence with the form of individual plants (e.g., leaf, needle, barrel trunk).[51] Grisebach himself extended Humboldt's enumeration of physiognomic plant forms to include fifty-four forms.[52] By incorporating new geological concepts, however, Grisebach began to pull Humboldt's concept of the regional physiognomy away from its Kantian origins. Grisebach's concept of formation sounded distinctly modern to later scientists and they frequently cited it, ignoring Humboldt's earlier contribution.

The geologists' concepts of "formations" and "succession" derived from their increasing understanding of geological strata and the distribution of fossils. As these concepts were formulated by Abraham Werner, William Smith, and Charles Lyell, they combined two distinct conceptions, categorical types and statistical intervals, that is, discontinuous types and continuous groupings, even though, of course, these seem inherently contradictory to our minds. This apparent contradiction produced an ambiguity in the conception of "formation" in the idealistic tradition, which was present also in Clements's later definition of formation.

At the end of the eighteenth century, Werner and Smith elaborated the basic laws of deposition of strata. Werner hypothesized that geological strata was formed by the deposition of suspended and dissolved particulates in a global, primeval ocean. The precipitates constituted a distinct series of "formations" recognizable by their mineralogical composition. His scheme allowed him to describe the geological column in terms of formations "succeeding" one another in the long process of deposition out of the ocean. William Smith demonstrated that strata contained distinct fossil assemblages. In the years from 1815 to 1817, he published a series of strata maps and explanatory memoirs that firmly established the utility of fossil analysis in determining the chronology of strata. Though he did not provide statistical frequency analysis of the varieties of fossils in strata, this was an easy inferential leap from his work and was made quickly in both Britain and France.[53]

In the mid-1810s, with fossil analysis of formations established, the outstanding question became the relationship between the fossil assemblages and the succession of formations. Did the series of fossils, from the deepest to the latest strata, exhibit any progressive trends of change? One of Charles Lyell's achievements, and, independently, of Gérard Paul Deshayes in France, was the perception of a statistical law of change in the successional distribution of the fossils. The implications of this discovery, that some fossils represented extinct species and that later fossils were of new species, wove themselves throughout Lyell's landmark work, *Principles of Geology* (three volumes, 1830–33), which so influenced Charles Darwin while traveling in South America. Regarding the statistical law of change, Lyell conceived of describing the fossil distribution in succession in terms of percentages of fossils in a formation identical with living species. His formulation appeared in the *Elements of Geology*:

> The result at which that naturalist [Deshayes] arrived was, that in the oldest tertiary deposits, such as those found near London and Paris, there were about 3-1/2 percent, of species of fossil shells identical with recent species; in the next, or middle tertiary period, to which certain strata on the Loire and Gironde, in France, belonged, about 17 per cent; and in the deposits of a third, or newer era, embracing those of the Subapennine hills, from 35 to 50 percent. In formations still more modern, some of which I had particularly studied in Sicily, where they attain a vast thickness and elevation above the sea, the number of species identical with those now living was from 90 to 95 percent.[54]

When combined with the succession of formations, the changing statistical distribution of fossils provided an intriguing synthesis of typological and continuum concepts. The formations were, of course, qualitative categories of geological phenomena, different to visual perception. Yet the changing proportionality of extinct fossils to living species was gradual, not changing in quantum leaps at strata boundaries. The changes in fossil distribution were detectable only quantitatively and were invisible to the eye.

The combination of typological and statistical frameworks for the analysis of succession of geological formations has a striking resemblance to the combination in Clements's ecological theory of typological framework for analyzing the succession of plant seres and the changing representation of plant species in the seres. No doubt this similarity was encouraged by the similar scientific situations of the men. Both were trying to find mathematical laws in phenomena first observed visually, that is, in qualitative terms. There is another source for the similarity, however; Lyell and Clements were both indebted to Humboldt's plant geography. Humboldt had combined the concept of qualitatively unique physiognomic regions of the earth with quantitative analysis of their habitat factors. Lyell drew explicitly on Humboldt's concept of the changing distribution of plants over the earth's surface. Humboldt showed that the number of species represented in a genus at the tropics is larger than the number of species represented in the same genus at the colder latitudes, because the warmer and wetter climates in the tropics permitted the greater proliferation of the genus's species. Assuming that the laws governing botanical distribution in the present era existed in the earlier geological eras having different climates, then it followed that statistical distributions for earlier eras should exist in analogy with the statistical distributions of the present. As the systematizer of geological uniformitarianism, Lyell presumed the laws of distribution and deposition were uniform across geological epochs. Therefore, the laws of the geological distribution of plants in the present era were related to the laws of historical distribution of plant fossils across geological eras.[55] As is well known to historians, this discovery was an important step in the intellectual development leading to Darwin's formulation of the theory of evolution. Clements later used a similar conceptual framework to relate the phenomena of contemporary succession to succession in geological eras. In *Plant Succession*, he tried to show that the law of succession explained the geological distribution of plant fossils. Clements depended in this, as had Lyell, upon a genre of quantitative analysis initiated by Humboldt.[56]

August Grisebach's synthesis of Humboldtian plant geography and historical geology in the concept of the plant formation was the basis for the work of his follower, Oscar Drude (1852–1933), professor of botany at the Dresden Botanical Garden. Drude had studied botany under Grisebach at Göttingen, and self-consciously considered himself to be carrying on Grisebach's work, after his teacher's death in 1879.[57] *Deutschlands Pflanzengeographie* and Drude's earlier *Handbuch* (1890), were the means by which the German tradition in plant geography, with its emphasis upon the integrity of the formation, was transmitted to Nebraska.

Drude emphasized the floristic formation, making it the basis of his geographical regionalization of plants. In the 1890 *Handbuch*, he listed five factors that determined the structure and extent of formations, with climate and rainfall playing as important roles as the form (e.g., leaf, needle) and social character (e.g., solitary, bunched) of the plant.[58] The climate provided the overall pull on the formation, determining its regionality, while the topography localized the associations of plants within the formation. "Wie das Klima in grossen Zügen, so gliedert die Topographie mit ihren Gefolge verschiedener Bodenklassen und Bewässerung im kleineren Massstabe die Formationen, macht aber eest das richtige Bild derselben fertig."[59] On the basis of these factors, Drude regionalized the flora of the world, and in passing described the Missouri prairie formation and other areas of midcontinental North America in terms that Pound and Clements were led to reject.[60] The Missouri prairie was treated as a unitary formation, and Drude—who had never visited the Americas—quoted from an 1845 botanical publication to say that it was dominated, as much as 75% to 90%, by buffalo grass (*Bouteloua oligostachya*, according to Drude). Pound and Clements thought they knew better; in moving west, the buffalo grass prairie did not begin until the center of Nebraska. Their effort to find the proper demarcation between the high grass formation and the buffalo grass formation led the two Nebraska scientists in the summer of 1897, as we have already seen, to invent the meter-plot.

More than Grisebach, Drude expressed a concept of succession: the formation developed, in a manner tied to climate and geological structure. Plant communities of past eras of the earth, as paleontological evidence certainly revealed, were transformed by changes in their condition. This history explained the contemporary distribution of plants and their systematic (formal) differences. "Die unformung der Pflanzenwelt durch Transmutation und die Umgestaltung ihrer Wohnorts—und Wanderungsbedingungen ist in Zusammenwirkung die Grundursach die gegenwärtigen Verteilungsweise bestimmter systematischen Sippen nach bestimmten Ländern."[61] Implicit in this scenario was the vision of one form of vegetation replacing another over time. The chief weakness of Drude's discussion of succession, and indeed, a weakness of succession in his whole tradition, was the failure to specify what happened biologically when one plant formation succeeded another. Understanding of the biological processes in succession would be a major contribution of the mechanistic tradition in plant geography.

Nothing in Drude's theory formally opposed a Darwinian interpretation of vegetation. Drude occasionally wrote of vegetational change in a Darwinian terminology; plant formations spread, invading other formations, and competed with them.[62] Drude's theory was essentially non-Darwinian, however, because he emphasized the social integration of the formation, making the formation behave as a unit in response to the climatic environment. The forest, prairie, moss, brush, freshwater, saltwater, and glacial formations had an ontological reality as units beyond that of the individual plant organism.[63] As a mechanist, Darwin tended to reduce group or social behavior to individual behavior, and vigorously expressed this reductionism in his analysis of the plant environment. Reductionism and opposition to the concept of the formation were hallmarks of the mechanistic tradition in geographical botany.

The Mechanistic Tradition in Plant Geography

Alphonse de Candolle was usually mentioned together with Humboldt, Robert Brown, the English botanist, and his father, Augustin de Candolle, as one of the originators of plant geography.[64] His *Géographie botanique* (1855) was a master work, influencing Charles Darwin, and was, in its own way, on the verge of the theory of evolution. Following in the footsteps both of Humboldt and of his father, Alphonse de Candolle stressed the importance of climate and past climates for the present distribution of plants and used a statistical methodology in his work.[65] He held a fundamentally different philosophical vision from Humboldt, however, of how nature worked. For Candolle, the plant was only a machine, responding to external stimuli. Here was no holism:

> En fait, une plante n'est point un instrument analogue au thermomètre, qui soit de nature à marcher parallèlement avec celui-ci; c'est plutôt une sorte de machine faisant un travail, et un travail très varié, sous l'impulsion des agents extérieurs, savoir, la chaleur et la lumière, et d'un agent intérieur, la vie, dont il est difficile de se passer pour rendre compte des phénomènes.[66]

The Swiss scientist emphasized the growth of the individual plant in its local environment, or station. The concern with habitat marked the mechanistic orientation of the scientist toward the plant, for in botanical usage, individual plants, but not formations, have habitats. This usage of station and habitat had been constant since the eighteenth century. Vegetation conceived broadly required equally broad causal agencies, of which climate was the only available candidate. As a scientist interested in precise determination of the causes of change in the vegetative world, de Candolle concentrated on the station of the plant, because this could, quite simply, be better known. Consequently, formations like the forest received less attention.

De Candolle criticized Humboldt's classification of the vegetative regions of the earth, seeing them as artificial and not supported by incontrovertible evidence. Humboldt's isotherm went through various definitions, but was never satisfactory. When isotherms were drawn in detail through a locality, they never precisely marked the floral boundaries as they were supposed to do.[67] There were no "natural" regions. De Candolle believed that only a statistical approach to vegetation was possible. Species could be counted and geographically plotted (a Humboldtian technique, of course), but floral statistics would correspond to physical factors only at the local level ("à des îles fort petites").[68] There were no *real* causal agencies to correspond to the region, as there were real causal agencies to influence the individual plant. The "physiognomy" that Humboldt had so poetically sketched (the perspectives of the Swiss Alps, Brazilian jungle, Argentine pampa) simply was not real. For de Candolle, the task of the botanist was to provide a mathematical description of the distribution of vegetation. Entities beyond the statistics, beyond the individual plant, did not exist.

Charles Darwin, who had been inspired as a youth by Humboldt's travels to become a scientist and whose own travel volume of the Beagle voyage followed the pattern of Humboldt's literary account, did not accept Humboldt's typological idealism. When after 1855, Darwin was writing the *Origin*, de Candolle's *Plant Geography* provided important evidence for his theory. While Darwin's theory of evolution and depiction of the natural world are too extensive to be reviewed here, I do wish to show how Darwin's discussion of plant geography opposed the concept of the formation, and the Humboldtian tradition of climatic causality.[69]

To have a Darwinian understanding of the functioning of the economy of nature, it is necessary to examine the individual organism and its relationship with other individual organisms. The biological interaction of organisms in the same food chain was the specific focus of evolutionary theory. The "relation of organism to organism in the struggle for life," Darwin wrote, was "the most important of all relations."[70] The individual was the unit of scientific analysis: not the group, not the species, not the formation. Only the individual was ontologically real. Darwin perceived the basic relationship between individuals as competitive. By definition, the Humboldt-Grisebach-Drude concept of a harmoniously functioning, ontologically independent group-unit, was impossible and false. Perception of formations was an illusion of the senses.

Darwin also jettisoned the notion of climatic causality. "I endeavored to show that the life of each species depends in a more important manner on the presence of other already defined organic forms, than on climate; and, therefore, that the really governing conditions of life do not graduate away quite insensibly like heat or moisture."[71] Darwin argued on behalf of an orientation toward life that we recognize today as ecological: the life of the organism is determined primarily and directly by the biological environment or relationships in which it is involved through food and reproduction struggles. Large causal factors like climate are filtered through the biological environment. The geographical boundaries of the biological habitat can be definite, because they are established by the limited boundaries of a powerful predator and will not "graduate away quite insensibly like heat or moisture." The broad parallelism between temperature and distribution of plant forms discovered by Humboldt cannot be discovered at the local level because climate is not a direct causal agency at the local level.

Precisely because Darwin so provided the intellectual framework for modern ecology, scientists have frequently supposed that the *Origin* was also the historical origin of the ecological work that immediately followed its publication. This is only partly true. Darwin's work initiated one line of ecological work, that of Eugene Warming, but left another (that of Humboldt) historically active, even if intellectually battered. Clements and the whole Nebraska school drew upon the Humboldtian tradition, as well as upon Darwin.

The Darwinian approach to vegetation was brilliantly applied by the Danish botanist, Eugene Warming (1841–1924). Warming was trained in the New Botany, in the laboratory of the distinguished Swiss botanist, Karl Wilhelm von Naegeli, at Munich. In the 1870s, he converted to Darwinism, for which he was an ardent spokesman the rest of his life. In 1895, he published *Plantesamfund*, translated into German and republished the following year as *Lehrbuch der ökologischen Pflanzengeographie; eine Einführung in die Kenntnis der Pflanzenvereine*. More than any other work in this period of the founding of ecology, it elaborated explicitly the concept of plant development, with a keen appreciation of the biological processes involved. It was written, moreover, without the benefit of Drude's *Deutschlands Pflanzengeographie*.[72]

Warming's separation from the idealistic tradition was obvious early in the *Lehrbuch*. He did not intend to use the term "formation." Its meaning was not well established and previous writers had used it loosely.[73] Presumably he included Drude in this charge, since he must have been familiar with Drude's *Handbuch* of 1890. Warming considered the plant "community" ("Pflanzenvereine") to be the basic analytical unit of plant geography. His concept of community was entirely Darwinian in meaning. The community was an association of plants in a small geographical area established by the competitive struggle for survival of the individual plants. The "bond" between plants in a community was called "commensalism."[74] Commensalism did not imply cooperation or harmony between plants in a community. Plants were not conscious entities and they were not integrated into a community in that sense. Commensalism arose as a result of every available biological place, capable of supporting life of some kind of plant, being occupied. If an occupying plant did not harm or destroy a neighboring plant (e.g., by depriving it of soil, moisture, or sunlight) in a Darwinian struggle, then it must have contributed positively to its survival. There was no question here of cooperation between plants to create a functioning community of ontological status "higher," or "more real," than that of the individuals themselves.[75] In the English translation of the book (1909), Warming endorsed the term "complementary association," as a synonym for commensalism. Association did carry a more appropriate meaning for English readers of the plant group's "bond."[76] The term "bond" carried intonations of a mystical or positive relationship between plants that Warming simply did not intend.

For many American scientists, the important points of Warming's work were his emphases on the struggle between plant communities and the development of plant communities through time. While he provided empirical information on the development of vegetative communities as examples, he also provided in Darwinian terms a biological theory of development. He related the theory of plant succession to the question of the origin of plant species. In analyzing these points, however, there are problems with the translations of Warming's book. The 1896 German language edition, which was the edition to influence H. C. Cowles and his colleagues at the University of Chicago, was rigorously reductionist, in the Darwinist sense, and implied no directionality to the historical development of plant communities. The 1909 English edition, which was also authorized by Warming, incorporated the new terminology of ecology invented by Clements. The Clementsian terminology imparted a faint flavor of teleology, which obscured Warming's rejection of Clements's monocausal (climatic) theory of formations. We can accurately characterize Warming's impact on American ecologists in the 1890s only by examining the German, and not the English, edition of his book.*

In the theoretical discussion of the struggle between plant communities, the German edition spoke adjectively of further devel-

---

*The German edition was, itself, a translation from the Danish, so it was possible that the Danish contained nuances of meaning obscured by the German. This is not a problem in the history of ecology, however, even though it might be a problem in a biography of Warming. All but one American scientist cited the German edition (only H. C. Cowles cited *Plantesamfund*).

opment of communities, while the later English edition said "successive." Thus, in a key section, describing vegetational development on new soil, Warming wrote in the German edition,

> Wenn irgendwo ein neuer Boden auftritt, so ird er bald von Pflanzen erobert werden. Es ist sehr auziehend, die weitere Entwicklung der Vegetation in allen ihren Phases zu verfolgen. Man wird Zeuge einer laugen Reihe von Kampfen zwischen den nacheinander einwandernden Arten werden; diese Kampfe werden bisweilen erst in vielen Jahrzehnten einen relativen Abschluss erreichen.[77]

In the English edition, the German "weitere" became rendered as "successive," and "Kampfen zwischen den nacheinander einwandernden Arten" was translated as "struggles among the successive immigrants."[78] To a casual inspection, the English translation conveys the same meaning as the German. But keeping in mind the distinction between the two nineteenth-century traditions in plant geography, on the one hand, and Clements's theory of succession, on the other, it is clear that the English conveys a directionality to plant development that was not present in the earlier edition. To the extent that Warming wittingly approved the translation of these phrases, he was introducing a Humboldtian nuance into his Darwinian theory. More likely, he was simply approving the Clementsian terminology of the English edition, without considering the extent to which that terminology falsified one of his central assumptions.

For Warming had made it clear in both the 1896 and 1909 editions that vegetational development was not directional and was not irreversible, both fundamental tenets of Clements's successional theory, and part of the reason that other English-speaking scientists, such as Cowles, spoke of vegetation as developing when they did not wish to endorse Clements's successionalism. For instance, in discussing the evolution of a lake into a heath, Warming wrote in the German edition, "Es ist nicht notwendig, dass die Entwicklung zuletzt gerade so, wie es eben angeführt wurde, vor sich geht."[79] In the 1909 edition, this passage, which spoke only of development, was translated to read "developmental succession."[80]

An essential principle of Clementsian theory was that every succession was headed toward a climatically caused monoclimax. Warming rejected the connotation of inevitability. In Clements's *Structure and Development of Vegetation* and *Research Methods*, a formation *had* to develop as the terminal climax of succession. For Warming, a plant community *might*, as it turned out, be the final outcome of development for a stable climatic region. The difference between the two formulations was one of tone, but tone was revealing of the different philosophical traditions standing behind each scientist. In both foreign editions of his work, Warming dealt very little with the notion of a final stage of development, going only so far as to state that when soil and climate were not changing, eventually one plant community (which in the 1909 edition was called a "formation") would "suppress all others" and "represent the *final stage* of development."[81] The verb "represent" was clearly intended to imply that "finality" was only relative. In absolute terms, there really was no final stage.

For readers of the 1896 German translation of Warming's work, therefore, the teleological nuances of the idealistic plant tradition were not present. Development was presented in a purely Darwinian framework. Both the individual plants in a plant community and the "community" or "association" in a habitat were in competitive struggle, the plants with each other and the associations with other associations in nearby habitats. This competition occurred for the Malthusian reason that each plant produced more seeds than the habitat could support; the seedlings competed against each other and against plants of other species for the available living conditions, in effect extending the territory of the species within the community. Eventually, if the soil and climatic conditions did not change, a stasis would be reached, in which the community would be dominated by certain species, with all available habitat niches occupied. Thus a forest would stabilize itself, with the dominant species, for example, a pine, thrusting sufficiently high to take all the light and growing densely enough to canopy the ground so that no other light-demanding species could penetrate. At the ground, shade plants, such as ferns and mosses, would provide soil cover.

The mechanics of individual plant competition within the association provided also for competition between communities. It is important to understand that Warming did not ascribe to the community itself an ontological reality separate from the plant, and hence the community did not have any mechanism of its own to compete with other communities. Rather, the same weapons used by plants in their internecine struggle were used by them also to invade neighboring plant associations. As Warming wrote, "The struggle between communities is of course dependent upon that between species."[82] By wind and animal dispersal of seeds or by extension of "root" systems, as with rhizomes, the boundaries of each plant community constantly pressed outward. This struggle is frequently observed in northern latitudes between meadows and forests. Trees invade fields. The trees normally possess sufficient advantages that a meadow will become a forest after a decade, if it is not cropped. The scene is familiar in deserted farmlands in the eastern United States. In some areas, such as the Great Plains, the lack of moisture prevents trees from establishing themselves naturally and grassland can choke out invading forests.

In one important chapter, "The Peopling of New Soil," Warming described the entire process of vegetational development. On new soil or denuded soil, as on new sand dunes, a few plants established themselves. So on Danubean river dunes, herbs first germinated, followed by rhizomes, which anchored the soil. Then "willows, poplars, alders, and other trees in the meanwhile grow up and produce bush-forest."[83] Sand dunes provided an important case of development. They usually were isolated as islands in the surrounding environment, either water or shore vegetation, and consequently could be exhaustively studied. Warming himself and other European investigators had studied dune development before H. C. Cowles published what became the classic American exposition of vegetative development in 1899.[84] Development ran from an initial community, usually of nonvascular plants, such as mosses and lichens, through transitional communities that anchored soil and prepared the way for dominant genera, such as vascular plants. When climatic and geological change occurred, such as drought and lowering of the water table or the opposite, the increase of rainfall, the dominant species changed and a new process of development began.[85]

The leading American student of Warming was Henry Chandler Cowles. Cowles had begun his study at the University of Chicago in geology under Thomas Chamberlain, the author of a popular textbook on geography and a specialist on tertiary period processes. When John Coulter came to Chicago to head the new Hull Botanical Laboratory, his famous gifts as a teacher drew Cowles into ecology.[86] In the summer of 1896, Cowles began a study of the shoreline of Lake Michigan, which he continued in the two following years.[87] In 1898, as a new doctorate and teacher at the Univesity, he brought his students for field excursions along the shore and also to North Manitou Island, Lake Michigan.[88] The product of these years of extensive study was his famous paper on the sand dunes of Lake Michigan. Cowles's choice of sand dunes was inspired directly by the work of Warming. Warming had not only devoted much space in the 1896 *Lehrbuch* to sand dunes, but had previously conducted famous floristic studies of Danish dunes. Cowles specifically credited Warming's dune studies.[89] The timing of Cowles's field trips was probably prompted by the German publication of the *Lehrbuch* of 1896. At all events, testimony was provided of Warming's influence by his appearance in the January, 1897, issue of the *Botanical Gazette*, published at Chicago, as an "Associate Editor." In a series of publications, Cowles presented the American version of Warming's ecological philosophy, intended, after 1904, specifically as an alternative to Clements's popular theory of succession.[90]

Cowles's affinity for the tradition passed to him by Warming, and his opposition to the tradition of which Clements was his contemporary advocate, were stated at the outset of his publications. In his study of sand dunes (1899), he defined "plant society" as "a group of plants living together in a common habitat and subjected to similar life conditions."[91] He specifically stated that the term was "taken to be the English equivalent of Warming's *Plantesamfund*, translated into the German as Pflanzenverein."[92] Plant societies were the result of specific causal agencies. The "formation," as used by Drude, had no similar value. "Formation" was a term applied to habitat groups considered together; it was a "generic" term of convenience and did not make reference to causal activity.[93] It did not have, therefore, the same ontological status as the plant society. Cowles similarly followed Warming in his rejection of the idea that the formation was the final stage of an inevitable development. In "The Physiographic Ecology of Chicago" (1901), Cowles took the epic perspective of Lyell and Darwin, that the climate and geological structure were always changing and stability was only apparent. In this framework, no vegetation could ever be stable. "In other words, the condition of equilibrium is never reached, and when we say that there is an approach to the mesophytic forest, we speak only roughly and approximately. As a matter of fact we have a variable approaching a variable rather than a constant."[94] The development of plant societies in wet climates toward the forest resulted in the forest as a culminating, rather than final, stage.

While Cowles believed that the climate had some broad causal relation to the vegetation, he did not agree that it caused the vegetation to develop toward a formation ideally suited to it. He did not agree that temperature, rather than humidity, was always the major component of a climate's influence. Following the mechanistic tradition of emphasis on the plant-habitat relationship, Cowles believed that the topographic physiography and the biological environment were the most important causal agencies acting on the plant and causing succession of plant societies.[95] Biotic causality was so rapid and powerful that it made climatic causality disappear: "So rapid is the action of the biotic factors that not only the climate, but even the topography may be regarded as static over large areas for a considerable length of time."[96] Of course, this had been the opinion of Darwin in the *Origin*. The immediate causality influencing plants and animals was the biological interaction created by the competitive struggle for survival; all other causality acted through this struggle.

By the end of the nineteenth century, two distinct approaches of explanation for vegetational change competed for advocates in the United States. One approach, which was to lead to Frederic Clements's mature work, *Plant Succession* (1916), was centered at the University of Nebraska and was the result of a formalization of the experience of Bessey's students by the theory of Oscar Drude and Clements's reading in sociology. The other was centered at the University of Chicago and represented by H. C. Cowles; it was stimulated by the work of Eugene Warming. The scientists in these two groups were well aware of the competitive nature of their different explanations to vegetation. Cowles, in a review of Pound's and Clements's *Phytogeography of Nebraska*, outlined the sociology of the situation, though he was not sure of the outcome of the competition:

> It may be too early as yet to predict whether the direction to future work in plant geography will be given by Warming or by Drude; and so whether we shall speak of ecology or phytogeography, or life forms or of vegetational forms, or plant societies or formations, is yet to be decided.[97]

As we have already seen, the Nebraska school's approach was deeply rooted in their experience on the prairies and in the Bessey instructional system. As we witness the tenacious Nebraskan loyalty to their vision of ecology, in the next section of this book, over two generations of scientists, we must always keep in mind that the difference between the Nebraskan and Chicagoan ecologies was not superficial, not a matter of terminology. They drew upon different European traditions of plant geography, were anchored in philosophically dissimilar visions of the natural world, and built upon many different, subtle nuances of feeling for and expression of change in nature.

5

THE LIFE CYCLE OF GRASSLAND ECOLOGY

Microparadigms for Plant Ecology

The appearance within twelve months of two alternative models of ecology, *Phytogeography of Nebraska* in 1898 and "Sand Dunes" in 1899, initiated a classic scientific competition. Would the grasslands be explored in terms of climatic formations, as in the Clementsian model, or in terms of topographical and biological habitats, as in the Warming-Cowles model? Scientists debated the merits in a wide spectrum of issues from philosophical to evidential, but resolution of the debate hinged on more than intellectual merit. The victory of the Clementsian microparadigm in grassland ecology depended ultimately upon the sociology of the competition: establishment of a regional hierarchy of teaching institutions, a network of research collaborators, interlocking research agencies, and the persistence—to return to the importance of individuals—of powerful friendships at the core of the advocacy of ideas. Stages in the history of the social structure of grassland ecology paralleled stages in the intellectual history of scientific ideas, from the outset of the competition to the victory and monopolistic utilization of Clementsian theory, and finally to the decline and abandonment of it in mid-twentieth century. In triumph, the tapestry of social institutions of Great Plains science supported an intellectual conservatism in ecological doctrine. Only the most profound events could unravel and fade the Nebraskan design: the tragic longevity of the midwestern drought, the biological revolution in genetics, which moved professional biology in new directions, and the aging and death of the premiere scientists.

The parallelism between the history of ideas and the history of the scientific community concerning the grasslands has been analyzed, on a theoretical level, by recent philosophers and sociologists of science. From their perspective, this parallelism results from the existence of a definite structure in the life cycle of scientific ideas and from the manner in which a scientific community becomes committed to those ideas. As a scientific paradigm for a specialty, or microparadigm, Clementsian theory followed this predictable structure. We will better understand the narrative history of grassland ecology if we are able to witness it through the framework of recent sociological scholarship.

The antipositivist philosophers of science, Michael Polanyi and Thomas Kuhn, argued that scientific knowledge did not accumulate as a simple aggregation of objective, empirical statements, collated by theories. Rather, scientific knowledge was intimately related to the personal commitments and tacit knowledge of the scientists. Intellectual advancement frequently came with wholesale abandonment of earlier scientific effort in a field and the adoption of new microparadigms. The sociologist of science, Diana Crane, recently tested certain aspects of the antipositivistic interpretation of scientific advance, by analyzing publication patterns of the specialty's literature. Crane assumed that a research field is established in a two-step process. First, the cognitive content of the field is created by an "innovation" that is initially confined to a small group of collaborators. The innovation can be theoretical or technical, such as the invention of a new apparatus and methodology. For the Nebraska ecologists, there were both kinds of innovations. Theoretically, their approach to ecology was based on the realization that the geographical and historical transitions between vegetative formations were continuous and invisible to sensory observation. The quadrat method of measuring interval variables was a technical innovation. Scientists at the University of Chicago did not have innovations of a similarly original character, but they were following a classically successful application of the Warming model of ecological analysis of the plant-habitat relationship, that is, H. C. Cowles's investigations of sand dunes. As we have seen, this model was built upon a philosophical approach to vegetation quite distinct from that of the Nebraska scientists. Both groups, the Sem. Bot. and the students around Cowles, largely pursued ecology in terms of their own model. Not until the publication of Clements's *Research Methods* in 1905 was the quadrat method widely adopted elsewhere.[1]

The second step in the establishment of the field is the diffusion of the innovation to scientists outside the initial collaborators. This diffusion takes place primarily by collaborative research and publication and by graduate instruction, that is, by personal contact. Of course, once the innovation is available publicly in journals, then scientists may choose to adopt it without any personal contact with the original collaborators. Nevertheless, most scientists so acquiring an innovation would likely seek contact with the originators, through correspondence, postdoctoral study with an originator, at conferences, or by any of the other means institutionalized in modern science. Thus, for instance, E. Lucy Braun attempted on her own initiative to adapt Clements's theory of plant succession to the eastern North American deciduous forests, but in advance of publication established a correspondence with Clements. Their exchange dealt with the full range of expected matters, such as interpretation of Clementsian theory, validity of its application to Braun's research, accuracy of the interpretation, and advice on publication. Thereby a network of collaborators is established, linked by the major figures in the field, such as Clements and his student, John Weaver. Crane called this network of linked collaborative groups an "invisible college." The network is "invisible" because generally it will not parallel the regular institutions of a discipline, such as professional societies, and because it is based on personal, even indirect, contact, rather than on contact formally initiated by the institutions.[2]

It is not necessary, of course, that the establishment of a research field should follow the two sociological steps outlined by Crane, or any number of steps proposed by any other sociologist. Indeed, one can easily think of research fields that were established with quite different histories. We are only now beginning to accumulate sufficient histories of specialties to understand the plurality of avenues of their emergence. Mendelian genetics is a famous instance of a research specialty whose initiator was unable to build up a group of collaborators, and whose innovation was "rediscovered" and spread by scientists in another field, horticulture, who had no contact with the originator. Nevertheless, some well-researched specialties have followed the two-stage scenario. Besides the examples of rural sociology and the theory of finite groups in algebra, brought forward by Crane, we can add that of radio astronomy, which has recently provided the most extensive case study of a specialty to date. David Edge and Michael Mulkay demonstrate that radio astronomy began in two small collaborative groups, at the University of Manchester and at Cambridge University following World War II, which formed around new apparatus and had low rates of publication for several years. Then a network was built, connecting groups of radio astronomers around the world.[3]

Each of the steps in the sociological emergence of the specialties has important intellectual correlates. The "innovation" represents a *dis*continuity in the intellectual tradition of a major specialty. The intellectual origin of the research area is not likely to be some point in the process of accumulation of empirical facts, as the positivistic interpreters of science have often maintained. The thesis that scientific revolutions are breaks with the traditional point of view of a discipline is well represented by Thomas Kuhn. Kuhn stated that the innovation was most likely to come as the result of the introduction of a perspective from outside the discipline, frequently from outside science proper, as Copernicus's astronomy was indebted to neoplatonism and Bohr's quantum theory was indebted to Kierkegaard's existentialism. While not subscribing to Kuhn's theory, Edge and Mulkay nevertheless perceive scientific innovation in radio astronomy as originated by scientists with interests in problems on the margins of the main discipline. They also comparatively note that marginal innovations by outsiders were responsible for new specialties in bacteriology, psychoanalysis, phage study, physical chemistry, and X-ray crystallography, which had been studied by other sociologists. In our history of grassland ecology, the Nebraska innovations of 1896 and 1897 clearly came from outside traditional plant geography and outside the variety of ecology being undertaken by the Danish scientists in the previous decade. It originated within the "Bessey system." This is similar to the history of radio astronomy, in which original workers were interested in practical problems of radar operation.[4]

Did the innovation come to constitute a microparadigm for the original collaborators? One of the key assumptions of Thomas Kuhn's theory, and of Diana Crane's variation of it, is that the innovation creating a new specialty is more than simply a technical solution to a problem. The innovation provides a novel orientation toward phenomena and a new methodology for solving a class of problems, or "puzzle-solving," as Kuhn put it. This microparadigm is contrary to the paradigm of the established discipline. Rejection of the microparadigm by the establishment leads to the new specialty and a competition between the old and new. It seems indubitable that the Nebraska innovations constituted a microparadigm for these scientists. Getting to the innovation required a mental switch, a new way of looking at the prairies; this constituted a nearly classic instance of what Kuhn calls the "gestalt-switch" behind a new microparadigm. Also, the innovation was and remained an orientation toward the world. It was not a logical or psychological triviality; as the subsequent history of the innovation will show, some scientists could never bring themselves to "see" the world the way the Clementsian ecologists did. When they challenged Clements, he was forced to reply simply that they had not had the experience they needed to understand the truth. If they asserted that the prairie was not an organismic formation, Clements would say, as in the privacy of a letter to a friend, that they simply had not *seen* the prairie. Finally, the innovation was an epistemological methodology. The quadrat method was a system of action, which guided participants, such as new students, to the vision of the prairie that underlay the specialty. Humans naturally tend to categorize phenomena in terms of sensory observations. Working the quadrat led the student away from that initial view of the world to the Clementsian view of the world. The quadrat method not only told the scientist "how to do" ecology, but also "how to believe" in the philosophy behind that ecology. This is why Clements adamantly insisted on the quadrat method of instruction. As Polanyi wisely understood, the master-apprentice relationship was central to learning a paradigm. Clements taught his students at Nebraska the uses of the quadrat without a textbook, by demonstration, for seven years before his *Research Methods for Ecology* appeared. Scientists—particularly European scientists—who did not utilize the quadrat were seldom able to accept Clements's vision.

Corresponding to the second sociological stage of the development of the specialty, according to Crane, is the intellectual period of science that Kuhn called "normal science." During normal science, the scientists with a particular microparadigm use their key innovation to solve "puzzles" spotlighted by their basic theory. We would expect, therefore, the Clementsian scientists to subsume more and more grassland phenomena under the schemata of successive development of the prairie formation outlined by Clements. This was "routine work," of the sort that graduate students undertake for doctoral projects. Consequently, as students were trained, they extended the "network" of the invisible college typical of this stage of the specialty's development. This stage of scientific advance resembles closely the description of science provided by Polanyi: tradition oriented; noncritical; closed to new ideas. The discipline was now intellectually mature. For Nebraska grassland ecology, this stage was built upon Clements's *Plant Succession*.[5]

Although publication is not the starting point of the relationships in the "invisible college," it is nonetheless the end point: the published dissemination of advances in knowledge. Building upon a suggestion of D. K. de Solla Price, Diana Crane has shown that the pattern of cumulative publications in a research area reveals the intellectual and sociological stages through which the specialization has gone. Crane's theory is built upon assumptions concerning the diffusion of innovations. In the first stage, in which the innovation originates and is largely confined to a small collaborative group, the growth in the publications concerning the innovation is slow. In the second stage of the invisible college, when a network of collaborative groups is being built, graduate training and adoption of the specialization increases the number of scientists practicing the specialty. Publications consequently increase in number. Crane has shown that the curve of cumulative publications in the specialization is logistic. During the first stage, the curve increases linearly; in the second stage, the curve increases exponentially. As the specialty's growth declines and ultimately ceases, the curve of cumulative publications correspondingly flattens out.[6]

Figure 2 shows the parallel sociological and intellectual stages of a Kuhnian interpretation of specialty development. According to Kuhn, "anomalies," which are counterinstances to the paradigm's basic theory, will accumulate and become the basis of another paradigm. If the new paradigm wins the ensuing competition, graduate training shifts to it, thereby shifting the pattern of publication away from the old paradigm. The old paradigm is abandoned. Crane has provided an alternative explanation for the decline of the specialty. She assumes that the old paradigm could simply be "exhausted." All the major problems that were the focus of the paradigm are solved and scientists' interest shifts to other fields. In this case also, the shape of the cumulative publications curve would be the same—a decline to a linear growth rate after the exponential growth rate of the normal science period, and then cessation. The decline of Clementsian grassland ecology involved both these explanations, and more.[7]

In figure 3, I have graphed the cumulative bibliography of grassland ecology, utilizing the Crane methodology. (For a full discussion

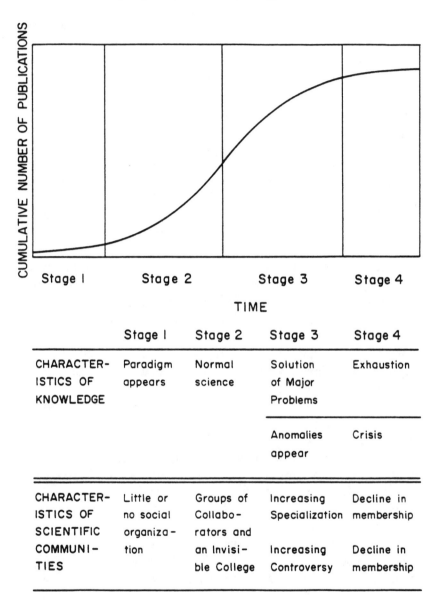

Fig. 2. Characteristics of scientific knowledge and of scientific communities at different stages of the logistics curve
SOURCE: Crane, 1972:172 (used with permission of Diana Crane and the University of Chicago Press)

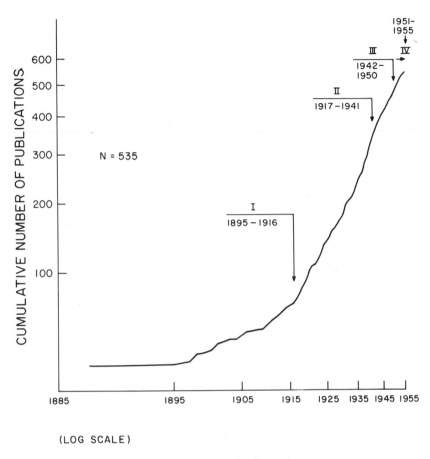

(LOG SCALE)

Fig. 3. Cumulative bibliography

of this application, see the Methodological Appendix, "The Grasslands Bibliography.") Chronological stages leap out of the graph. The years between 1895 and 1916 constitute the founding period of the specialty, when scientists wrote the basic literature, competitive microparadigms were offered, and when, toward the end of the stage, the victorious microparadigm was chosen. After 1916, the normal science period exploded with publications, rapid increase of the practitioners of the specialty, and establishment of the research and training networks. The normal science period ended in 1941, after which decline of the grassland group was evident in publication rate, training, employment, and (though we could not determine this from the statistics alone) intellectual innovation.

The pattern of grassland publications reveals the sociological structure underlying its intellectual development. The authorships of papers are oriented by allegiance to the competing microparadigms of stage one. In stage two, as one microparadigm wins this competition, one set of authorships withdraws from the literature, leaving the publications and the intellectual development of the specialty in the hands of the followers of the dominant paradigm. The institutional bases of the collaborative networks reveal themselves in the biographies of the authors. Analysis of the publication pattern of grassland ecology spotlights the separate groups of scientists competing to establish their microparadigm in the early history of the specialty.

Underlying Kuhn's and Crane's theory of the emergence of a specialty is the assumption that the lives of scientists reflect the development of ideas. This assumption denotes a normative, or rule-guided, interpretation of science, and is contrary to the assumption of some sociologists that specialization occurs for primarily sociological reasons, for example, that an established discipline is already fully stocked with researchers and their institutions, so that younger scientists can obtain recognition only by setting up a new field. We have already seen, in the previous chapters, that the Nebraska scientists were not primarily motivated by need for status recognition or competition with an established discipline. Their motivations grew out of their student admiration of Bessey and the social enthusiasm of the laboratory milieu he created. The formation of the Nebraska botanical group and several years of scientific research preceded its professional status. When professional recognition finally did come, after the 1898 publication of the *Phytogeography of Nebraska*, a dozen scientists had entered and left the Sem. Bot., displaying in their movement a variety of motivations, but demonstrating little need to be known for their contributions to a new specialty. Recall that of the 1890 Sem. Bot. group, Pound went into the law, Marshland taught high school, Stoughton left botany for the ministry, Webber became a federal citrus researcher. When the Sem. Bot. took on a professional role in Nebraska in 1892, by initiating the botanical survey, Rydberg was only a graduate student and Clements was an undergraduate. Their lives exhibited a richness and diversity of situation and intent that should caution us against explaining their interests in a new specialty solely by a need to establish a new field for careers or to obtain status recognition. Without doubt, they wished to conduct their scientific research in a professional way, guided by the watchful supervision of the rigorous Bessey, but that did not mean that they were only influenced by the professionalism of science. They were not, to use a phrase that was once fashionable, one-dimensional men. The scientific lives of the members of the group were guided by the scientific ideology conceived within the Bessey system. They inculcated Bessey's values concerning scientific method and the role of scientific knowledge in a democratic society; they shared the life experience of the prairies and the intellectual vision, after 1897, of continuous change in the grassland formations; after the invention of the quadrat, they shared indoctrination in an exciting new method and its implied epistemology.[8]

Educational Preparation of Grassland Scientists

The collective biography of the grassland scientists reveals the patterns of microparadigm competition through educational preparation, employment, and research and collaboration networks. The subsequent domination of the normal science period by Clementsian scientists appears in cessation of the training in competitive microparadigms, the growth of a Clementsian "establishment" in regional research, and the preponderance of citations in the literature to Clements and his colleagues.

In the granting of doctoral degrees, the University of Nebraska and the University of Chicago dominated grassland ecology from 1895 to 1955 (see table 3). On the one hand, this might not seem surprising, since both were major institutions in the Midwest. If we consider that there were other important universities in the region that failed to contribute scientists to the specialty, however, the extent of the dominance is clearer. The state universities of Kansas and Iowa and George Washington University in St. Louis were situated on the prairies and might have been expected to become centers of doctoral instruction in grassland ecology, but they did not. The University of Minnesota produced three ecologists, and was involved in the network of Nebraska grassland scientists. The two primary doctoral institutions were the seat of the two microparadigms for the doing of grassland ecology. Here was a one-to-one

Table 3
Degree Origins

Degree origins	B.S.	M.S.	Ph.D.	Total no. of degrees (B.S., M.S., Ph.D.)
Univ. of Nebraska, Lincoln	14	18	24	56
Univ. of Chicago	2	3	8	13
Kansas St. Coll., Fort Hays	8	6	0	14
Univ. of Minnesota	1	1	3	5
All others (1 or 2 each)	29	15	11	55
(No data or no degree)	(4)	(15)	(12)	

correspondence between the intellectual and institutional homes of the specialty.

While doctoral instruction was focused in two universities, baccalaureate preparation was dispersed over the Midwest. Only Nebraska and Kansas State College, Fort Hays, produced a large number of undergraduates who went into the specialty. Again, we would expect Nebraska to take this role. Fort Hays's role as a feeder institution came from its involvement in the research network of the Nebraska microparadigm. Grassland ecologists came from such different undergraduate institutions as Utah State College, the University of California, the University of Cincinnati, Butler University, and Purdue University. The pattern of dispersion of undergraduate institutions and concentration of the doctoral institutions indicates that the specialty was truly regionally based, with recruitment for the graduate programs following a structure, rather than being accidental. While we do not, of course, know the motivations of the undergraduate students who decided to travel to Nebraska or Chicago for graduate work, we have no reason to doubt that these were the familiar motivations of going to the institutions with reputations, based largely on research publications, for the specialty.

The chronology of awards of educational degrees (App. fig. 3) has little that is unexpected in it. If we examine degree award in the third stage, however, we do find a clue to the unexpected rise in publications in 1950. Note that the last baccalaureate was awarded in 1942, but seven doctorates were awarded between 1947 and 1952. Recruitment into the specialty from the undergraduate level had ended in the third stage, yet professional training continued at a strong pace. It is not difficult to conjecture that the strong finish of doctoral production in the late 1940s was a result of the "postwar" veteran returning to finish his education. Indeed, the mean age of the doctorate in the third stage was 38.33 years (see fig. 4), by far the highest of any period. Interesting as are these aspects of education, however, they do not specifically relate, as far as I can tell, to the microparadigm.

The chronology of academic origin of doctorates strongly reflects the microparadigm (see fig. 4). In the first stage of the specialty (1895–1916), there was an equal production of doctorates by Nebraska and Chicago, with five each. In the second period, 1917–41, the University of Chicago gradually withdrew from the production of doctorates in grassland ecology, while degrees awarded by the University of Nebraska increased. Nebraska trained twelve scientists and Chicago prepared three. This was a dramatic cessation of doctoral competition in this field between the two schools, and undoubtedly must relate to important institutional decisions by the University of Chicago. While I have not been able to obtain primary material to explain the institutional decisions at Chicago, circumstantial evidence leads me to believe that the university simply shifted its interests into other specialties. After 1939, only the University of Nebraska trained scientists who sustained the grassland specialty.

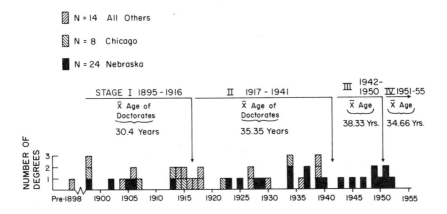

Fig. 4. Academic origin of doctorates, 1895–1955

Divergent Missions in Education and Research

The University of Nebraska and the University of Chicago, as the two leading graduate centers in grassland ecology, presented divergent orientations toward education and research.

For two generations, from 1884 to 1955, the study of the midcontinent grasslands at the University of Nebraska was led by a succession of teachers trained in the Bessey system. Bessey himself taught at Nebraska until 1913, retiring shortly before his death in 1915. His student, Frederic Clements, taught there from 1898 to 1907. John Weaver, who was a major figure in the normal science period, was an undergraduate at Nebraska from 1905 to 1909, and took his master's degree from Nebraska in 1911. Then he followed Clements to Minnesota, where Clements had gone in 1907 to replace Conway Macmillan, who had retired to private business. Macmillan had been trained under Bessey at Nebraska, receiving his M.A. (his highest degree) there in 1886. Weaver received his doctorate at this Nebraskan outpost in Minnesota in 1916. In 1915, he returned to Nebraska as a teacher in the botany department, where he remained for the next forty years. Clements left the University of Minnesota in 1917 to join the Carnegie Institution (in Tucson), which had committed itself to research in ecology. Although he remained with the Carnegie Institution, conducting research in alpine ecology in Colorado in the summers and coastal ecology in California, Clements never severed his ties with the University of Nebraska. His Carnegie-sponsored alpine laboratory at Manitou Springs, Colorado, was a summer research center for Nebraska graduate students from before World War I through the 1920s. Certainly, this was a tight interlacing of student-teacher tradition!

Missing from the Nebraska tradition after 1907 was Roscoe Pound. In 1898, he began teaching at the Nebraska law school, which required him to curtail extensive botanizing. He remained in Lincoln, Nebraska, of course, and undoubtedly remained a close friend of Bessey and Clements. The record book of the Sem. Bot., which had been kept by Pound and others, ended in 1899, though the seminar itself went on. Pound's educational law career brought increasing success and eventually pulled him away from Lincoln. He was dean of the law department at Nebraska from 1903 to 1907. He jumped to Northwestern University to teach law in 1907, then moved to the University of Chicago in 1909. Finally, in 1910, he joined the law faculty of Harvard University, where his brilliant career as dean and theorist of jurisprudence reached its culmination. As long as he was in Lincoln, he tried to maintain some botanical activity, as far as I am able to tell without extensive primary evidence. He supervised the botanical survey of the state until 1903, and consulted with Clements on *Research Methods*, as we know from Clements's acknowledgments. His irrevocable departure in 1907, however, removed a central mind from the Nebraska group.

After Weaver's arrival in Lincoln in 1915, the University of Nebraska had a stable group of grassland researchers who provided leadership through the late 1940s. Besides Weaver, the botany department included Raymond J. Pool, who was its chairman for several decades. Pool was entirely a Nebraska product, receiving there his A.B., M.A., and Ph.D., with a lone year (1908) passed at the University of Chicago. Located in a neighboring department was Franklin Keim, chairman of agronomy, who had received his bachelor and master's degrees from Nebraska, but sojourned in Ithaca, New York, long enough to take his Ph.D. from Cornell University. Dr. George E. Condra was head of the Department of Geography and Conservation. Condra was a contemporary of Bessey and Clements and active in the 1920s and 1930s. He was extraordinarily vital, heading various Nebraska state commissions in the drought decade. As late as 1950, he published an article on Nebraska geological history. No doubt the long-term strength of the University of Nebraska in grassland studies derived from the generational stewardship of these scientists. The interdisciplinary character of the group, including the fields of botany, geography, conservation, and agronomy, was typical of the Nebraskan approach to the prairies. At the same time, associated with the agricultural station at the university, their work always carried an indirect utility to plains agriculture. The group embodied the Bessey system.

If the University of Nebraska fulfilled the mission of the land-grant university, the University of Chicago epitomized, by comparison, the ethic of the ivory tower. The dedication of the university to pure research, unsullied by immediate problem solving, has been a legendary chapter in the history of American higher education. Since the University of Chicago provided the main competition to

Nebraska in preparing doctoral scientists to study the grasslands, especially before 1916, I devote a brief discussion to the institutional context of the education of Chicago scientists.

William Rainey Harper directed the new university, in the 1890s, toward " 'pure' scholarship and science."[9] Applied studies, in the sense of vocational training or research on problems brought from outside disciplinary science, were far from the university's purpose. The ideology of the first president was perpetuated by his successors. "Research should be the outstanding feature of the University of Chicago," third president Ernest Burton told the university senate in 1923.[10] Sixth president Lawrence Kimpton expressed the university's image of itself in 1953 with a rehearsal of its history:

> We were the ones to announce in the heartland of American practicality that the distinguishing mark of a university is research. We went further than this. We said that no matter how others might come to use or abuse the word, by "research" we meant the pursuit of fundamental knowledge. And we said, secondly, that we at least should be concerned with the search for new knowledge for its own sake. From the beginning, we were fully aware that the essence of university research is to be completely free and uncommitted to practical results.[11]

The university's Department of Botany did not deviate from the founder's line. John Coulter, the first chairman, came out of a church school background; he taught at Wabash College, which was Presbyterian, and presided over Lake Forest College, also Presbyterian, before coming to the university. He was close to the natural history tradition that looked back to John Ray, who found evidences of God manifested in his creation; he was not oriented toward the German university. His sense of the relationship between intellectual inquiry and man's problems was similar to President Harper's: science saved men's *souls*. According to Coulter's biographer, Andrew Denny Rodgers III, Coulter believed that pure science in botany was the special province of the university. The federal government and agricultural experiment stations adequately took care of practical botanical matters. What the Midwest needed, Coulter felt, was a great department of pure botany. This vision was apparently behind his early involvement with the University of Chicago while he was still president of Lake Forest College.[12]

Coulter did believe in a service ethic, specially defined, as part of science. In a book in which he was trying to reconcile evolutionary and religious approaches to life, he wrote:

> The ideals of science may be summarized as follows: They are first, to understand Nature, that the boundaries of human knowledge may be extended, and man may live in an ever-widening horizon; second, to apply this knowledge to the service of man, that his life may be fuller of opportunity; and third, to use the method of science in training man, so that he may solve his problems and not be their victim.[13]

This ethic was not utilitarian. Coulter did not expect ordinary individuals to bring scientists their everyday problems to solve, which was precisely the land-grant ethic placed on Bessey at Nebraska. It was expected there that farmers would write or bring in their problems, and this situation did lead to a new understanding of the grassland for the Nebraska group. Rather, for Coulter, science developed in directions determined by its own internal intellectual concerns. Eventually, scientific progress would yield practical benefits. Coulter's view was central to the ideology of science in the 1920s and was shared by other nationally prominent scientists, such as the Nobel prize-winning physicist, Robert Millikan, who was for a long time Coulter's colleague at the University of Chicago.[14]

The university's antipathy to the research orientation of the state universities was explicit. The decision from the first had been to distinguish the University of Chicago from the state universities. In 1923, President Burton pointedly drew the comparisons:

> This brings me to my fourth and final reason for emphasizing research, viz. that very few other universities are in a position to give to it the first place to the extent that we are. I speak not in derogation of other institutions and for your ears only, but it is evident that a State University with its obligations to give a collegiate or technical training to the increasing numbers of the youth of the State, who are demanding it, responsive as it must be to the opinion and demands of a constituency to whom research means little or nothing, compelled to seek the money for its support from State legislatures in annual or biennial sessions, may indeed undertake research, but will find it practically impossible to give to it the place of first importance.[15]

The approach of Chicago botanists to the new specialty of ecology reflected the overall orientation of their university and the self-consciously created differences between it and the state universities. Chicago's ecology was concerned less with controlling than with preserving the natural world. Clementsian dynamic ecology grew out of and returned for nourishment to the practical soil of agriculture. The choices at each institution of first fields for ecological study revealed the differences. H. C. Cowles chose to study sand dunes. Dunes were apparently selected because they were previously studied by Eugene Warming, whose work decisively influenced Cowles and because that work offered a disciplinary base from which Cowles could analyze dune succession. Cowles thereby came by his research subject through a purely academic route. Pound's and Clements's choice to study prairies came out of the botanical survey of the state, which had in its turn been undertaken for utilitarian considerations. Choices of sand dunes and grasslands were not simply geographical convenience. Illinois contained extensive grasslands and true prairie remnants, which were studied later with a characteristic perspective of pure research. Nebraska contained sand dunes that sprawled over several counties. Bessey, Pound, and Clements all visited and studied them, but they did not study them in the 1890s from the point of view of Warming's *Oecology*. Rather, the Nebraskans wanted to reforest the sand hills and bring them under control, an interest that carried into the next generation of Nebraska botanists.[16]

As the new specialty of ecology established itself in the 1890s and 1900s, Chicago botanists expanded their geographical areas of study, while Nebraskans remained essentially at home. Chicago botany had been well provisioned by a million-dollar endowment in 1896. A large, four-story laboratory building was built (The Hull Botanical Laboratory). The Chicago facilities were the best in the Midwest.[17] Almost immediately, the Chicago shop was larger than Nebraska's. Within two years, Chicago had produced a half-dozen professional botanists for the scarce teaching appointments, including ones at Syracuse University, Smith College, the University of Rochester, the Shaw School of Botany (St. Louis), and McMaster University, Toronto.[18] Of course, Cowles, who received his doctoral degree under Coulter in 1898, was teaching at Chicago and was "in charge" of plant ecology.[19] With these facilities, a large staff, and a large number of students, Chicagoans were able to travel across the continent. In the ten years from 1898 to 1907, Cowles took his students to

North Manitou Island, Lake Michigan, the Tennessee mountains, the Flathead Lake area of Montana, Woods Hole, Long Island, Mt. Katahdin, Maine, the Gulf of Mississippi, Vermont, the deserts of Arizona, the Florida Everglades, Oregon, and Alaska.[20] In these same years, Nebraska students ranged no further from home, that I have discovered, than the Colorado Rocky Mountains, where they studied with Clements in summers.[21]

Chicago botanical ecologists frequently worked closely with the geography and zoology departments of the university. Nebraskan interdepartmental relations, as we have seen, tended to be with the applied sciences, like agronomy, and the experiment station. Cowles worked closely with Charles B. Davenport, a zoologist, eugenicist, and the author of a pioneering American statistics textbook, discussed in chapter 3. Davenport teamed with Cowles to lead the ecological expedition to the Gulf of Mississippi in 1902.[22] The zoology department became, indeed, an early producer in zoological ecology. Victor Shelford, who received his doctorate in zoology in 1907, adopted a Clementsian scheme in the study of marine ecology, and later, while at the University of Illinois, contributed to grasslands ecology. (Shelford was a member of the Grasslands Research Federation and also chairman of the Committee on the Ecology of the Grasslands of the National Research Council.)

The study of ecology at Chicago became sufficiently prestigious that in 1913, a British ecologist, visiting Chicago as part of the International Phytogeographical Excursion of 1913, could refer to the "Chicago school of ecology."[23] Nevertheless, Chicago did not produce a coherent group of scientists working on grassland ecology, although eight of their doctorates contributed three or more articles to the specialty. The eight Chicago scientists were (with the date of their doctorate): Stanley Cain (1930), David F. Costello (1934), Henry Chandler Cowles (1898), George D. Fuller (1913), Paul Sears (1922), Victor Shelford (1907), Arthur G. Vestal (1915), and Stephen Sargent Visher (1914). In addition, although he did not take a doctorate, John H. Schaffner studied for a year at the University of Chicago after his master's degree (1896–97). That these scientists were not a research group was indicated simply by the lack of collaboration. Of course, this was strikingly different from Nebraska, where Clements and Weaver both collaborated widely and frequently.

Employment Patterns

The sociological maturity of a research specialty manifests itself in the cohesion of the scientists' careers and their collaborative networks. Kuhn and Crane assumed that the sociological structure of the research community would be influenced by the microparadigm. Put somewhat differently: the career patterns of those scientists aligning themselves with and sustaining a microparadigm should be different from the career patterns of scientists rejecting that set of ideas. We must expect therefore that the career patterns of the Nebraska grassland ecologists would have been significantly different from those of the Chicago scientists.

Turning first to employment over careers, we see a striking contrast in the variety of employers of the two groups of scientists. (See figs. 5, 6, and 7.) The scientists of the University of Nebraska had significant, long-term involvement with four major groups of employers, all colleges and agricultural stations as one group; federal research agencies, secondary and normal schools, and "all others," a category including private and philanthropic institutions, such as

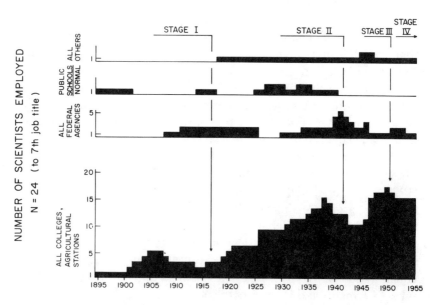

Fig. 5. Employment patterns of University of Nebraska doctorates

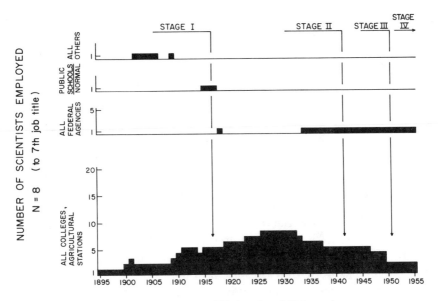

Fig. 6. Employment patterns of University of Chicago doctorates

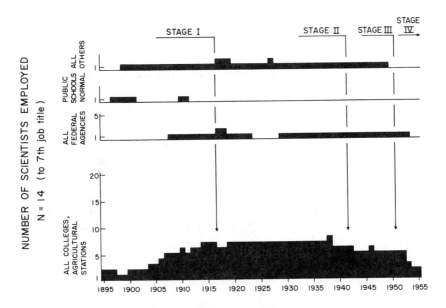

Fig. 7. Employment patterns of all other doctorates as a group

endowed botanic gardens and the Carnegie Institution. In contrast, scientists of the University of Chicago restricted themselves, with few exceptions, to working in universities. Questions obviously arise concerning the assurance with which we ascribe these differences in career patterns to acceptance and rejection of the Clementsian micro-paradigm. On the one hand, the patterns meet the expectations of the Kuhn and Crane theories, and also can be related to the character of the Clementsian ideas. On the other hand, we can perceive the differences as owing to the character of the universities and their prestige levels. While I would not deny that these are contrary explanations, I think a closer examination of them will show that they are parts of a broader explanation.

The Clementsian microparadigm for plant ecology was distinguished by its strict adherence to the principle of causality in plant-habitat relations and in succession of seres in formations. It was intimately associated with agricultural technology. Clements derived much of his instrumentation for analyzing habitat factors in the quadrat from horticulture. And the motivation for his mathematical analysis of habitat factors was simply to discover the variable responsible for plant adaptation and associational structure in the habitat. His major publications other than *Plant Succession* had direct relevance to economic ecology.[24]

A microparadigm that took so much from agriculture gave much in return. The Nebraska school tackled numerous ecological problems of agriculture, from early encounters with Russian thistle in the 1890s to the protracted struggle with drought in the 1930s. Graduate students were introduced to the agricultural concerns and thesis topics often derived from grassland economic problems, in the sense of looking for the scientific problem in the practical problems of farmers. It is not fair to say that these thesis problems were not therefore scientific; they were derived from a disciplinary base. The interest in the problem was generated by the Nebraskan milieu. The Nebraska graduate students were well prepared by their training in the Clementsian microparadigm for employment in experiment stations and in federal research agencies, such as the forest service, which had midwestern research stations, and the Department of Agriculture. We should expect that the government would want to hire such ideally trained scientists. So Nebraska doctorates worked for the Forest Service, Soil Conservation Service, Soil Erosion Service, Division of Botany and Agrostology—USDA, and the Bureau of Plant Industry—USDA.

Similarly, the long interest of Bessey in secondary education legitimated his scientists' employment in high schools and normal schools. Although the scientific curriculum in high schools was weakened by the efficiency and industrial education movements of the years preceding the First World War, a few Nebraska doctorates turned to secondary school employment in the mid-1920s, when research budgets were cut (see fig. 5). There was nothing incompatible between professional research and public school employment, even though it might be difficult to conduct research while so employed.

The University of Chicago had its own definition of the university's research mission. The botanists and ecologists at Hull Laboratory did not have the involvement with the practical problems of midwestern agriculture that Nebraska scientists had. Their ecology certainly had economic applications, should they care to make them, for instance, in arresting the spread of lakeshore dunes; but generally the interest of Chicago scientists grew out of their microparadigm candidate, rather than out of concern for the problems of agriculture. This is not to say that their handling of their problems was more "scientific" than the Nebraskan scientists; indeed, two of the Chicago scientists (Shelford and Costello) adopted the Clementsian microparadigm.

If the experience of Chicago graduate students with "practical" problems was narrower than that of Nebraskan scientists, their field experience was wider. We have already seen that Cowles journeyed summers to Florida, Alaska, the Appalachian Mountains, and abroad. Precisely this wide experience made them valuable to academic departments of botany and biology. There is little meaning to the question whether this experience made them *more* valuable to academic departments than Nebraska scientists with their different experience, but certainly departments interested in training graduate students broadly in ecology without a necessary orientation toward economics would find a Chicago scientist an important addition. Only one Chicago scientist (David Costello) had any federal employment experience and he held the Clementsian microparadigm

(Costello was chief of the Division of Range Research of the Rocky Mountain Forest and Range Experiment Station of the Forest Service).

We might be tempted to explain the difference in career patterns between the two schools' scientists in terms of their different prestige levels. While there is no way to disprove this possibility—I do not have access to the motivations of interview committees—several considerations lead away from it. During the first period of the specialty's history, the Nebraska botanical faculty was as eminent as the Chicago botanical faculty. Bessey and Coulter were scientific peers, and before Clements left Nebraska in 1907, he had established a reputation as the leading innovator in ecological methodology. In the 1920s, the general prestige of Nebraskan botany was not as high as under Bessey, but it remained the leading scientific department of Nebraska. Chicago lost Coulter in 1926, when he retired to head the Thompson Institute on Long Island, leaving behind a distinguished department, but one without the luster of a founder of the New Botany. Botany was under reorganization in Chicago, and would be combined into the biology department, indicating that the university believed that plant science would be better advanced within a broader framework. Furthermore, Nebraska scientists taught at the leading midwestern colleges and universities, as did their Chicago counterparts. Nebraskans taught at Nebraska Wesleyan, Iowa State College (Ames), Kansas State College (Fort Hays), Kansas State College (Manhattan), Montana State College, University of Oklahoma, Baker University, the University of Minnesota, University of Nebraska, University of Minnesota (Duluth), Catholic University (Washington, D.C.), Saint Cloud State College, University of Montana, Northwest Missouri State College, York College, University of Nevada (Reno), University of Illinois, Texas Agricultural and Mechanical College, Southern Illinois University, and Michigan State College.

Further evidence that the Nebraska scientists worked where they did for professional reasons, rather than reasons of prestige level of the university, is provided by the patterns of research centers in grassland ecology. Nebraska scientists tended to work at colleges and universities where grassland ecology was a long-term concern, and this should be expected, since they published in the specialty. This is easily perceived when we group the scientists according to college employer. Of the twenty-four Nebraska doctorates who published in grassland ecology, thirteen worked in nine centers of grassland research, with nine in the two largest centers, the University of Nebraska and Kansas State College, Fort Hays (App. table 4).

Some of the research centers in grassland ecology shared a feature that reinforced the connection between academic ecology, plant physiology, and agriculture: the agricultural experiment station. Nebraska, Fort Hays, Berkeley, Ames, and North Dakota were associated with experiment stations. When we consider that the experiment station system was partially funded by the federal government, we see immediately a tightly interlocking network of academic and federal research institutions, including the stations and the USDA research agencies. No need exists for a prestige-differential explanation for the difference between the career patterns of Nebraska-trained scientists and Chicago-trained scientists.

Finally, the chronology of employment in the career patterns corresponds closely to the production of the literature of the field. The exponential explosion of grassland publications in the 1920s and 1930s was produced by a rising tide of employed Nebraskan scientists. The number of Nebraska doctorates employed in colleges and experiment stations increased from five in 1920 to fifteen in 1938, while the number of Nebraska scientists employed by the federal research agencies increased from one in 1930 to five in 1940 (see fig. 5). This increase was pushed on by the great drought of the 1930s, which prompted an increased federal budget in the USDA for research and mitigation related to the drought. State universities similarly increased their staffs. The Clementsian microparadigm, as we shall see in the final chapter of this book, was employed to analyze the drought cycle and the effect of drought on vegetation. Scientists trained in this useful microparadigm were in demand by the agencies expected to deal with the crisis.

Collaborative Networks

The network of collaboration in research and publication was an important aspect of the careers of the grassland scientists. Superimposed on the network of research centers, with scientists trained from Nebraska and Chicago, was a pattern of coauthorships which revealed the continual postdoctoral flow of research between the centers. One aspect of the sociological theory of normal science is that collaboration is a major means of expanding the number of scientists using a microparadigm. Coauthorships are, of course, only one indication of collaboration. Much collaboration would not reach the stage of coauthoring articles, and would be revealed by published acknowledgments and, indeed, only by private correspondence. Nevertheless, coauthorships are public identifications of the highest level of collaboration and hence serve to award credit and prestige. I focus on coauthorships, without excluding the idea that further collaboration might be revealed by other means.[25]

As predicted by Crane, the number of coauthorships rose rapidly in the normal science period (see App. fig. 4). The rise and decline of the coauthorships closely follow the publication rate of the grasslands bibliography as a whole. The University of Nebraska and federal scientists produced the greatest number of the coauthorships (60% of all coauthorships), and the University of Chicago scientists coauthored *no* grassland publications. Listed by school of doctoral origin, coauthors showed a strong orientation to the University of Nebraska (see App. table 5).

When the coauthorships are graphically illustrated (see fig. 8), the relationship between the major and minor figures in the field is startling. This configuration is precisely what Crane has meant by an "invisible college." She writes that "members of these [research] groups were not so much linked to each other directly but were linked to each other indirectly through these highly influential members."[26]

The size of the grasslands collaborative network was typical of specialty networks in general. In rural sociology, only eight scientists were high producers, with more than ten publications, and in the theory of finite groups (the mathematical specialty Crane examined), only four mathematicians were high producers, with more than ten publications. Comparative statistics are provided in table 4 on the productivity grouping of the specialties. Examining figure 8 on the largest multiauthorship complex, we can immediately see that the four high producers were the central figures for the coauthorship patterns. And we can see that the moderate and low producers of the literature tended to be tied to the major producers rather than to each

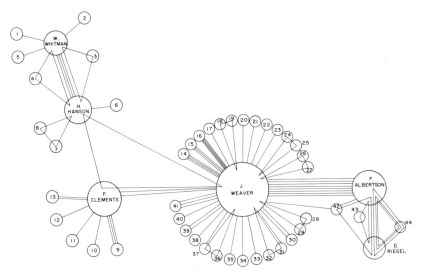

Fig. 8. Multiple author relationships: Clements-Weaver complex (80 multiauthor
publications, representing 42% of all multiauthor publications and 15% of
all publications in grassland bibliography)

Interpretation:
Lines terminating at circles couple coauthors (see App. table 6 for key)
Lines angling through circles link three joint authors (spatial configuration
for clarity only.)

other directly. The only exception to this was the Costello-McIlvain-
Savage complex of moderate producers in the federal service.

When the geographical location of the high producers is noted, the
collaborative network is seen to reinforce the grassland research
centers network of graduate degree origin. Frederic Clements and
John Weaver were located at the University of Nebraska (Clements
having left in 1907), Albertson conducted his entire career at Kansas
State College, Fort Hays, and Hanson was located at the North
Dakota Agricultural College. In other words, all the features of the
"invisible college" research network reinforced each other: research
centers in the specialty took their scientists from the institutions
identified with the microparadigm; the high producers were located
at the research centers; the scientists with many coauthored publications were located at the research centers and tied peripheral scien-

Table 4
Productivity Grouping

| | Producers | | |
Specialty	High (more than 10 pub.)	Moderate (four to ten)	Low (less than four)
Grassland ecology	4	25	29
Rural sociology	8	11	33
Finite groups	4	13	47

Source: Rural sociology and finite groups, from Crane, p. 49.

tists, not located at these centers, into the network through their coauthorships.

In contrast to the Nebraskan invisible college of grasslands ecology, the University of Chicago, with eight doctoral scientists, had no mutual collaborations and none of these scientists was a high producer. Four of the Chicago scientists, however, were moderate producers in the field. We can conjecture that the lack of a collaborative network of Chicago scientists was probably owing to the failure of a Chicago scientist to become a high producer in the field. The key scientist at the University of Chicago who might have so welded a network, but did not, was H. C. Cowles. As a teacher and later department head at Chicago, he was in an ideal position to have challenged Nebraska in grassland ecology. The reasons that he did not do so are not public. But we have already seen the beginnings of an institutional explanation for the University of Chicago's failure to build up a research network in grassland ecology. By the early 1920s, the methodology and the basic theory of the field were widely conceded to be Clementsian. The University of Chicago undoubtedly wished to put its research interests into fields to which it could make major contributions. The Clementsian paradigm came under criticism in the late 1920s for its failure to incorporate population genetics in its theory. By the mid-1920s, the future of the theory of evolution lay in genetics and the task for biologists was to synthesize genetics, natural selection, and ecology. The University of Chicago moved in the direction of this synthesis and the Nebraska grasslands network did not do so. The strength of the collaborative network that enabled Nebraska to dominate grassland ecology was a vested interest so huge that the field could not move to a contrary biological paradigm.

The final feature of the collaborations in grassland ecology that I wish to point out is the existence of a small collaborative network of federal scientists, separate from the University of Nebraska network (fig. 9). None of the major scientists of this federal complex, David Costello, Ernest McIlvain, and David Savage, had any educational connection with the University of Nebraska. Indeed, Costello had been trained by Cowles at the end of that great teacher's career. The separation of this coauthor complex implies a federal literature within the grasslands bibliography. Separation does not imply isolation, however; indeed, Costello, who tied the federal complex together, was a Clementsian scientist. His intellectual allegiance to the Clementsian paradigm was a final indication of the extent to which Clementsian ecology came to dominate the specialty. In his recent evocation of *The Prairie World*, Costello acknowledged his "greatest debt" for "inspiration and knowledge" to four friends: Fred Albertson, F. Clements, H. L. Shantz of the Forest Service, and John Weaver.[27]

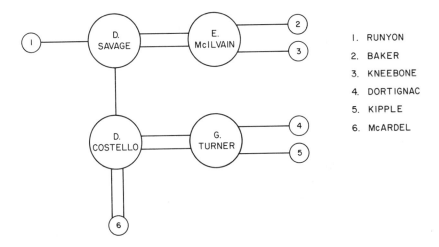

Fig. 9. Multiple author relationships: Costello-Savage complex

Known by the Company They Keep: Citing Behavior

The detailed parallelism between the intellectual and sociological growth of grassland ecology implies that the way in which the scientists *used* scientific literature was also definitely structured (as distinguished from the way in which they *published* their literature). In contemporary science studies, sociologists analyze the patterns of citations in publications to discover the research networks, prestige ranking, competitive groupings, and other social attributes of a specialty. These studies have frequently been successful and the results not necessarily obvious. Even though the grassland bibliography began nearly sixty years before the data analyzed in contemporary studies, the citation patterns of this much older field are similar to those of contemporary specialties. As historians, we have the advantage of seeing the citation statistics over the full life history of the specialty. In this longer perspective, the citation pattern of grassland literature, dominated monopolistically by Clementsian theory in the normal science period, reflects the specialty's social structure. Use of literature relates to employer and generations of literatures (that is, of different groups of titles) rise and fall like the tides over the lifetime of Clementsianism. The citation statistics reinforce the image, obtained in the previous sections of this chapter, of a tightly interwoven, inward-looking, conservative specialty. These statistics may seem dry as dust, as abstract as an index of the stock market, but we must remember that the scientists making them, penning them on sheets of paper in drafts of articles, usually knew each other personally, worked in the same research network, and felt the same prairie in their hearts.

The fifty-eight scientists of the grassland school made 5,717 citations in 210 publications. (See the Methodological Appendix for a discussion of the methodology regarding the citation analysis.) Of these citations, fully 1,932, comprising over 33 percent, were to themselves. Each scientist was cited, at the median average, 18 times. This median value may surprise us as low; nevertheless, the figure is significant in the specialty. It provides the possibility that each scientist's works were cited by 18 of their 210 publications (more than 8 percent). Also, this median count is high, when compared to the median aggregate citation count of nonsubject scientists, which is

3 citations. Considering that one-third of the total citations were to themselves, a median of 18 citations clearly does not preclude their domination of bibliographic referencing. The scientists at the center of the coauthorship and research networks were most frequently cited: John E. Weaver, 679 aggregate citations; Fred Clements, 252; Fred Albertson, 178; Herbert C. Hanson, 85 (the Methodological Appendix, table 7, provides a ranked listing of their aggregate citation counts). The sociologist, McCann, suggests that the list of citations provides a prestige ranking of scientists, on the assumption that recognition is the primary reward in science. The most-cited scientists garner the greatest prestige mainly by their productivity and intellectual contributions. Without entering the thicket of controversial assumptions we must make to see the cumulative citation listing as a prestige ranking, we at least observe that the most-cited scientists also turn out to have been the major contributors in the narrative account of the specialty's history.[28]

The low median value of aggregate citations to the works of nonsubject scientists implies that only a few nonsubject scientists would have aggregate citations as high as the median of the subject scientists. Nineteen nonsubject scientists accumulated at least 14 citations to their works, an aggregate far below both the median and modal aggregate citation counts of subject scientists. These nonsubject scientists specialized frequently in ecology, but also in tangential fields, such as agronomy and systematic botany. They were not isolated from the subject scientists, however; some of them with high aggregate citation counts were coauthors with the subject scientists. Fitzpatrick was the coauthor with John Weaver of the most-cited title in the bibliography, which accounted for 40 of the 60 citations to Fitzpatrick's works. Pound was, of course, coauthor with Clements of three major works, accounting for 30 of his 33 citations, including the landmark, *Phytogeography of Nebraska*. All of W. Hansen's citations were as a coauthor with Herbert Hanson. Russel and Smith were not linked to the subject scientists by coauthorship. (See the Methodological Appendix, table 8, for a ranked listing of nonsubject scientist citations.)

From the pattern of the aggregate citations, I conclude that the fifty-eight subject scientists were primarily a self-referencing group of researchers. This conclusion is important as a raw symptom of the Kuhnian postulate of the isolation of normal science. As we can see in chapter 8, the grassland ecologists were isolated from external critics, and external criticism apparently did not contribute to the theoretical collapse of the specialty. Although one internal critic, H. A. Gleason, was highly cited, the major external critics, A. G. Tansley of Oxford University and the plant sociologists of the Braun-Blanquet school, were not. Indeed, Braun-Blanquet's important *Plant Sociology* was only cited seven times. I do not claim that the low level of citation of external critics proves that external critics were not listened to; after all, who wants to cite a critic frequently? And who knows how many criticisms were silently absorbed? My point is simply, of course, that absence of citations to critics is a symptom of talking among friends, whom one does not expect to demur or to raise a critic's standard against your theory.[29]

As an aggregate statistic, citation totals for the scientists cannot indicate much about citation behavior. It is likely, for instance, that different publications were cited at different rates. A scientist whose works carried a low aggregate citation count might, nevertheless, have had a higher rate of citation; that is, his works might have been more frequently used. I have chosen to provide rates of citation in several steps. In the first step, presented here, I examine gross rates of citation to all of a scientist's works considered as a whole. These gross rates provide us with a sense of levels of activity in the specialty, in analogy to the impression we would have if we were a scientist reading the journal literature year by year. Certain scientists' names appear and reappear, impressing us, as one among many indicators, as influential. In a self-referencing community, such as grassland ecology, the gross rate of citation ought to distinguish between scientists at the research core and those at the margins. Gross rate of citation may be used with information about collaborative networks to make such judgments. The problem with the gross citation rates is that they do not pinpoint influential publications. This is crucial, since citations are not of a scientist's name, but to his or her specific titles. In a second step of analysis, I develop a set of normalized rates of citation for the fifty-four most frequently cited titles in the specialty's history. These statistics provide a view of generations of grassland literature.

Scientists who were not among the fifty-eight subject scientists or associated with them through coauthorship dropped in importance in the gross rate ranking, even though their aggregate citations may have been higher than the median or modal values for subject scientists. (See the Methodological Appendix, tables 9, 10.) Subject scientists whose aggregate citations may have been below the median value nevertheless had a higher gross rate of citation than nonsubject scientists. Thus J. Russel and J. Smith, nonsubject scientists largely not affiliated by coauthorship, who had 26 citations each, had gross rates of only 0.67 and 0.41 citations per year respectively (App. table 10), while E. H. McIlvain, with only 15 citations, had a gross rate of 2.14 citations per year (App. table 9). These results reinforce the conclusion presented above that subject scientists did not as frequently cite nonsubject scientists as they did themselves, and further confirm the picture of the subject scientists as a self-referencing research community.

Assessment of a scientist's influence must be based on the period of the specialty's history to which he contributed. We would not judge two scientists to be of equal influence, on the basis of gross rates of citation, if we knew that one scientist's rate had been earned with publications dating from the founding period of the specialty and the other scientist's rate located solely at the closing stage. Appendix table 11 sorts out the scientists according to the stages of the specialty in which their first cited title appeared and ranks them within stages according to gross rates of citation. Immediately, we see which scientists of the preparadigm period had lasting influence. Clements, whose first title was published in 1892, and Weaver, whose first cited title appeared in 1915, were easily the most important scientists. Notice that H. C. Cowles, who offered a preparadigm theory to compete with Clements's, ranked low among the founders in the references of grassland scientists.

Despite the sorting of the scientists by stages of growth, the suspicion may nevertheless linger that gross rate of citation is a simple result of the age of the cited titles. Thus, older publications could have higher rates than recent publications, because the older the title, the more opportunity for citation, and, citations attracting further citation, the higher the rate of citation. Alternatively, more recent publications could have higher rates, because scientists tend to cite a research front, quickly retiring older publications; as the older publications increase in number, each older title is cited less often, since scientists are unwilling to increase the percentage of their references to older titles. As it turns out, neither scenario is accurate.

In general theoretical terms, Derek Price has demonstrated that the age of publications is unrelated to their rates of citation. Older titles are sorted and cited in a complicated manner, he believes, related to cumulative advantage. As for recent literature, as the pace of publication has increased in this century, the recent literature available for citation by a scientist at any time has been so huge that most of it is not cited at all. In the instance of the grassland publications, the correlation coefficients are low for the relationship between date of publication and gross rate of citation. For all scientists in all periods, the correlation coefficient (Pearson's r) between the date of publication and the gross rate of citation is $+ 0.13$; for scientists appearing in Stage I, $+ 0.19$; for scientists of Stage II, $+ 0.24$; for scientists of Stage III, $- 0.24$.[30]

Generations of Grassland Literature

Thomas Kuhn's suggestion that changes in the patterns of citations should reveal revolutions in science undoubtedly did not refer to patterns of citations to authors, but to individual titles. We generally expect that scientists cite authors because of what they publish specifically, rather than who they are (though pandering goes on in science, certainly, as elsewhere). Let us turn, therefore, to citations to titles. I have found that the most frequently cited titles fell into four generations.* These were generations of use, in which groups of titles were highly used for their period, then fell into comparatively less frequent use, being replaced by another group of titles in high usage. Each literary generation held a topical unity, as well as chronological unity. The first generation of titles was, of course, comprised by the founding works of ecology, for instance, Warming's *Plant Ecology*, Schimper's *Geography of Plants*, Cowles's "Physiographic Ecology of Chicago," Pound's and Clements's *Phytogeography of Nebraska*, and Clements's *Research Methods in Ecology*. References to these titles dropped drastically after 1928. Indeed, Warming (referring to *Plant Ecology*) was cited only twice after 1928, Schimper only once, and Cowles not at all. Clements's *Plant Succession* was chrono-

---

*For a discussion of the methodology by which the most highly cited titles were located, see the Methodological Appendix. Generational titles are listed in the Methodological Appendix, tables 12, 13, 14, 15, and 16.

logically within this founding generation of titles, since it was published in 1916; but it received its highest use within the next generation of titles, the normal science group (and has been listed with them). The normal science titles represent a unity of intellectual approach to the grasslands. They are dominated by citations to *Plant Succession* and Weaver's and Clements's *Plant Ecology*, the leading textbook for the specialty. With twenty-four highly cited titles, the normal science literature was the largest generation of titles and clearly represented the bulk of the specialty's research work. I have divided this literature into two sections or generations, Normal Science A and Normal Science B. Normal Science A is comprised by normal science titles published before 1928, but which, nevertheless, have their highest rates of citation after 1928. The normal science literature had its major usage in the period between 1929 and 1949, though titles were preparing for it from 1916 to 1928. Normal Science B comprises titles published after 1928.

The Great Drought of the Midwest of 1933−41 erupted in the middle of the normal science period of the specialty, producing its own literature. The drought titles are chronologically the most tightly bound. The literature began only with the appearance of the phenomenon and terminated sharply in 1946. The mean rate of citation of drought titles from 1934 to 1946 was more than three times the mean rate of citation of the same titles after 1946.

The final generation of titles was symptomatic of the erosion of grassland ecology as a scientific specialty and its transformation into an intellectual technology. After 1946, range management emerged as a strong topic within grassland literature, and by the 1950s range management researchers undoubtedly outnumbered scientific researchers, with range management dominating publications. Range management, as a profession, was largely a product of the Great Drought, and range management literature had a straightforward intellectual relationship to problems raised by the drought. But the general relationship between the range management titles and the drought titles is rather complicated, for the scientists most immediately connected with solving management problems created by the drought *under*cited the drought publications.

The periodicity of the four generations of grassland literature corresponds roughly to the sociological stages of growth and decline for the grassland scientists. According to the pattern described in figure 3, the upward slope of the normal science period began following 1916 (the publication date of Clements's *Plant Succession*). We do not see such a sharp break in citation patterns at 1916; rather, the founding literature was replaced distinctly after 1928 by a new generation of titles, which had been emerging since 1916. The year 1929 saw a new edition of Clements's masterwork, and also witnessed the publication of Weaver's and Clements's *Plant Ecology*. To the extent that these publication events revealed their authors' intuitive sense of the intellectual change of the profession and its readiness for textbooks, they also reveal the year in which a trend of slow growth of a normal literature surfaced into high-use visibility. Without any exceptions, the titles of the founding literature fell into disuse after 1928.

The transition from a preparadigm literature to a normal science literature based on a microparadigm seems, in summary, to be a gradual matter of the replacement of one literature by another, rather than a sharp change, as might be expected from examination of the growth curve of the specialty's bibliography. The decline of the founding literature and emergence of the normal science literature correspond roughly to the decline of the University of Chicago's graduate training in grassland ecology and the increase of the University of Nebraska's training. Coupled with the increased Nebraska interest was the emergence of a regional research network in grassland ecology, centered around Nebraska, but also including federal research centers and several smaller independent centers.

The final generational shift in grassland literature coincides with the change from Stage III to Stage IV of the specialty's sociological growth, as described by figure 3. The years between 1951 and 1955 correspond with the emergence of range management literature citations of this subject from 1950 to 1955. Although normal science citation continued in these years, it finally halted in 1956, when no titles were listed for grasslands ecology in *Biological Abstracts*, but titles in range management were listed. Many of the scientists who had earlier contributed to the scientific specialty were still active in 1955, of course, but the orientation of the specialty had so decidedly shifted that we may say that the specialty had ended by the mid-1950s despite their presence. Many of them were undoubtedly contributing to range management; some were at the verge of retirement. Clements's *Plant Succession*, which had been annually cited, often two to four times, before 1950, was cited only once after 1950. Weaver's and Clements's *Plant Ecology*, similarly heavily cited before

1950, was cited twice in 1951 and not again thereafter. In contrast, revealing the applied economic interest of range management, Clements's *Plant Indicators*, which had important uses in agricultural research, was frequently cited in Stage IV.

Scientists who are familiar with the scientific literature of the post-World War II years would say that in the late 1940s, the Clementsian paradigm was under attack. By reference to the citations without simultaneous reference to the intellectual history of Clementsianism in this discussion, I do not want to appear to deny the intellectual challenges to it. The citations simply indicate a shift in generations of literature. A shift out of the normal science period could be explained by challenges to Clementsianism, but I believe the evidence will show that the disintegration of normal grasslands science resulted from internal disillusionment with the microparadigm, rather than external attack. The presence of the generation of drought literature and the emergence of range management within normal science are indicators of internal disruption in grasslands ecology, but any explanation of the causes of this disruption must rest upon the content of the literature and the lives of the scientists, which are examined in the final section of this book.[31]

The presence of range management literature and normal science literature together in the final years of the grasslands specialty naturally raises the possibility that two separate groups of scientists had emerged, one concerned with theoretical or paradigm-oriented questions of the specialty, and one concerned with economic problems. Of course, economic problems had stimulated the theoretical interest of Bessey and his students at the end of the nineteenth century, and these two concerns were never far apart in the history of the specialty. Nevertheless, in the normal science period, the Clementsian paradigm provided a large constellation of problems to be solved that referred directly to theory and only indirectly to economic problems. I have assumed that academic scientists would turn to paradigmatic grasslands ecology, with their theoretical labors expected to produce indirectly utilitarian benefits. Ultimate practical benefit to flow from "basic research" is an assumption in the ideology of pure science. Federal scientists concerned with grasslands were located in the Department of Agriculture, where they had a mandated responsibility to economic problems. By placing academic scientists and federal scientists into two categories, I do not mean to intimate that their interests were opposed, simply that their orientations within grasslands ecology were different and might reasonably have led them to use differently the generations of literature.

To test whether citation rates differ for employment categories of scientists making the references, I have determined the rates of citation for the generations of literature in different historic periods for three groups of scientists, academic (including agricultural stations), federal, and "all others" (including scientists in botanic gardens or employed by private, noninstructional research institutions like the Carnegie Institution). The data from this test are displayed in Appendix table 17. These data are normalized for both years and number of scientists making the references. The generations of literature have been discussed above. The historic periods begin with the Founding Literature (1895–1916). Although *Plant Succession* was published in 1916 and initiates the normal science period, I have chosen to deal with the year in two distinct ways. For the Founding Literature, I have ended the founding period in 1916, rather than 1915, to allow citations in 1916 before *Plant Succession* was read. I have begun Normal Science A with 1916, rather than 1917, however, since Clements's book was published that year. Thus Normal Science A runs from 1916 through 1928, and Normal Science B runs from 1929 through 1949. The Drought Literature period was from 1935 to 1946. Range Management was from 1947 to 1955. The period of the normal science decline was between 1950 and 1955. Since these statistics are normalized for both the number of years of each period and for the number of scientists doing the referencing in each period, rates of citation may be compared with one another.

Examination of Appendix table 17 permits an empirical scaling of the citation rates, with a normal range including scientists of all employers. In this scale, 0.01–0.07 (citations/in-print year/manpower year) (n = 8) represents the empirical low range of citation rates; 0.08–0.14 (n = 13) is the middle or normal range; 0.15–0.32 (n = 6) is the high range. The rate 0.00 (n = 8) is not included in the scale, since it indicates that a literary generation was not being cited at all, rather than being cited at a low rate. Rates above 0.32 (n = 1) are extraordinary. This empirical scaling is, of course, not theoretically derived, so we have no basis for believing that the citation rates should or should not have fallen in the ranges of it that they did. The scale only allows us to make normative judgments; that, for instance, when federal scientists cited Normal Science Literature B at a rate of

0.14 in the years between 1950 and 1955, they were citing at an expected rate in terms of their own behavior and the behavior of academic scientists. Also, when federal scientists cite at 0.14 and academic scientists cite at a rate of 0.09, they are both within the expected range and we ought not to judge their rates as significantly different.

The Founding Literature

The Founding Literature was used within the normal range by both academic and federal scientists until 1928, after which date citation of the literature nearly ceased. Scientists employed in other institutions tended to underuse the founding literature, but in the long years from 1917 to 1928, even they utilized it with expected frequency.

The rapid abandonment of the founding literature after 1928 should be expected for grassland scientists, who had a new textbook for grassland ecology in Weaver's and Clements's *Plant Ecology*. There is every reason for scientists to turn to textbooks specifically for their field and away from general foundational works, once a normal science period of the specialty has been achieved. Indeed, the textbook is typical of normal science, according to Kuhn. We can also see the citation of Warming's, Schimper's, Cowles's and the other titles of the founding generation for about two decades as a long life, since Price had found that the average scientific publication is regularly cited only for ten years. It may well be that these founding works were cited for such a long period simply because no textbook was available until 1929.[32]

Table 5
Citation Rates of the Founding Literature
(Citations/In-print Year/Manpower Year)

Employer	1896–1916	1917–28	1929–49	1950–55
Academic	0.13	0.09	0.03	0.02
Federal	0.10	0.14	0.00	0.00
Other	0.03	0.09	0.02	0.00

Normal Science Literature A

While academic and federal scientists tended to cite the Founding Literature at much the same rates, their citations of Normal Science A were distinctly different. In the years between 1916 and

Table 6
Citation Rates of Normal Science Literature A
(Citations/In-print Year/Manpower Year)

Employer	1916−28	1929−49	1950−55
Academic	0.10	0.11	0.32
Federal	0.29	0.08	0.03
Other	0.05	0.08	0.00

1928, federal scientists cited Normal Science A at an extraordinary rate, while academic scientists cited it at a normal rate. At the end of the history of the specialty, however, the situation was quite reversed. Academic scientists used the generation of titles at an extraordinarily high rate in the years between 1950 and 1955, while federal scientists nearly ceased citing it at all. The statistics make clear that federal scientists had virtually withdrawn from normal science, in the Kuhnian sense of the term, after 1949. This withdrawal, however, took place in two steps. Although the reasons for the extraordinary usage of Normal Science A before 1929 cannot be determined from the statistics, after that date, federal scientists decreased their use by 72 percent. Almost certainly, we must look to broad institutional changes to have forced this shift. The decline in federal use of Normal Science A in the 1950−55 period was of a similar magnitude, 63 percent less than the previous period (1929−49). The institutional reasons for this latter decline have already been mentioned, that is, the creation of a range management literature and the shift of federal scientists, in line with their mandated obligations, to it.

Normal Science Literature B

Both federal and academic scientists used Normal Science Literatures A and B with an expected or middle range rate of citation in the years between 1929 and 1949. And as with Normal Science A, academic and federal scientists used Normal Science B differently in the final years of the specialty (1950−55). Federal scientists increased their rate of citation of Normal Science B in the 1950−55 period by 75 percent over the previous period, but remained within the middle range of citation. Academic scientists, in contrast, increased their rate of citation in the 1950−55 period by 246 percent over the previous period, pushing it into the high range of citation.

Table 7
Citation Rates of Normal Science Literature B
(Citations/In-print Year/Manpower Year)

Employer	1929–49	1950–55
Academic	0.13	0.32
Federal	0.08	0.14
Other	0.03	0.00

Federal scientists, unlike academic scientists, used normal science titles at different rates in the years between 1950 and 1955. If A titles were virtually uncited, B titles were within the same range of citation as federal citation of A and B titles in the previous period.

What may we conclude from the uses of A and B titles? We may first confirm our earlier inference that the retirement of a literature does not depend on its age, but upon what scientists use it for what reasons. Federal scientists retired an earlier literature (Normal Science A) in the 1950–55 period, but academic scientists increased their use of it. Derek Price has shown that the retirement of cited titles is not simply a result of established patterns of use, that is, a less-used literature is not necessarily retired before a more-used literature. He conjectures, without support, that the retirement is related to the aging of the scientists. I think that we can advance beyond that suggestion. In the broadest terms, citation behavior is related to the knowledge-making enterprise of the specialty. Ecologists cite ecologists; ecologists do not cite sociologists, per se. Within these broad categories of usage, citation behavior is modified, in terms familiar to sociologists of science, by status level of scientists being cited, or whether a scientist is in a strategic research center, and so on. It had not been clear, however, whether these modifications of citation behavior are significant in historic terms, that is, over generations of literature. The statistics I bring forward in this chapter indicate that one enduring determinant of citation behavior is the intellectual interest of the scientists in the literature. The differences between academic and federal scientists' usages of Normal Sciences Literatures A and B are plausibly explained in this way. As is well known to ecologists themselves, Clementsian theory came under heavy attack (and ultimate rejection) after World War II. In the course of this heightened theoretical interest, we would expect academic scientists to cite the literature more frequently than federal scientists. Federal scientists, however, were not primarily interested in the theoretical qualities of Clementsianism, since their concerns had shifted toward range management. As a result, they did not take part—we should conclude from the citation behavior—in the theoretical controversy, did not use the Clementsian or theoretical literature unrelated to economic problems, and consequently undercited the normal science literatures. At the same time, it is not likely that the federal scientists failed to cite the normal sciences literatures by reason of not being linked to the central research institutions or most prestigious scientists of the normal science period, because many of the federal scientists were linked to the University of Nebraska at some point in their education or were involved in the Clements-Weaver-Albertson-Hanson coauthorship network. The citation rates for the years between 1947 and 1955 are compatible with this interpretation.

Drought Literature, 1935–46

Citation rates isolate the scientific literature of the great midwestern drought of 1933–41 as a special group of titles in the midst of the normal science generation. This literature was distinguished by its close attention to the effects of the drought on the vegetation of the plains. Most of the titles were authored by Weaver and his doctoral student, Fred Albertson, who was located at Kansas State College, Fort Hays, one of the long-term research centers in grassland ecology. As it turns out, Weaver and Albertson were also its most frequent citors. Because of the peculiar insular quality to the "drought industry," as one Weaver student called it several decades later, academic scientists cited the drought titles at a rate five times higher than the federal scientists. The federal scientists were, indeed, no more interested in the drought publications than they were Normal Sciences A and B during this period. Academic scientists, in contrast, cited the Drought Literature at a rate three times the rate of their citation of the Normal Science Literatures.

Table 8
Citation Rates of Drought Literature
(Citations/In-print Year/Manpower Year)

Employer	1935–46	1947–55
Academic	0.55	0.24
Federal	0.11	0.03

Abandonment of the Drought Literature was similarly differenti-ated by employers. College scientists dropped their interest in the drought after 1946 with a 56 percent decline in citation rate. Federal scientists turned away even more sharply, their 1947–55 rate of citation of Drought Literature was 73 percent lower than the 1935–46 rate of citation. In other words, academic scientists' interest after 1946 dropped from extraordinary to normal, while federal scientists' interest declined from normal to negligible.

It should be surprising that federal scientists did not develop a high interest, at least as revealed by citations, in the Drought Literature. Obviously, the drought was of great interest to the federal govern-ment in the 1930s. The number of federal scientists increased in that period, as we saw in the previous chapter in examining the career patterns of Nebraska graduates alone. The Department of Agri-culture was reorganized to accomplish its research mission better. A presidentially appointed national committee investigated the drought, for which our grassland scientists provided expert testimony, as we shall see in chapter seven. Federal programs to mitigate the effects of the drought are textbook knowledge in American history and a living memory of many Americans. Despite the intense institutional focus on the drought, federal scientists did not *highly* cite drought publica-tions. The reason for this cannot be determined from the citation statistics, but several interpretive suggestions can be offered. First, the major mission of the federal research agencies was not scholarly, and many federal publications were in a popular format. Federal demonstration work, field station research, and extension activity were aimed at enabling the farmer and rancher to increase produc-tion or farm more profitably. Publications frequently did not carry the full scholarly apparatus of notes and bibliography. The drought undoubtedly intensified the utilitarian approach of the federal scien-tists. Their research mission was not to examine fundamental theory of grassland ecology or agronomy, rather to apply what they knew or to develop ad hoc technologies to mitigate the effects of the drought. Only if the theory dramatically broke down in attempts to apply it might we expect federal scientists to have turned to basic research in this emergency. But the practical applications of Clementsian ecol-ogy worked well in the drought. Theoretical postulates of the paradigm were, indeed, imperiled in the drought, but these did not immediately affect applications.

Academic scientists might, however, have found their interest in

Clementsianism stimulated by the drought. It provided a geographically enormous and temporally long test of the theoretical assumptions of the theory, such as the inevitable progressivism of vegetative change on the prairies. John Weaver, as Clements's chief disciple, was particularly led to examine the drought as a test of his master's theory, since it struck at his notion of the ability of the prairies to "repair" themselves. The citation patterns by themselves do not, of course, provide any basis for deciding if these were the reasons that an academic interest in the Drought Literature was greater than interest in Normal Sciences A and B, but a reading of the literature itself and archival research in the institutions make it the most plausible explanation.

A second reason that federal scientists might not have been as interested in the drought in a scholarly way as were academic scientists relates to their definition of their own work. Following World War II, the Range Management Association was formed, with largely federal scientist membership. The drought of the preceding decade was apparently responsible for bringing together the various federal research groups, for example, plant pathologists, grassland ecologists, agronomists, into a single professional context, range management. It is possible that the drought shifted the federal scientists' self-definition of their research mission in such a way that they were less interested in paradigm-oriented drought literature. While the citations provide no evidence that this was the case for the years between 1935 and 1946, they do provide evidence of high federal interest in range management following 1946.

Range Management Literature, 1947–55

Federal scientists cited these titles at a rate 667 percent higher than their citation of the Drought Literature. Academic scientists, true to their paradigm-oriented interests, did not cite the range management titles significantly more often than the Drought Literature and for a comparable (but not identical) period, cited them 19 percent less than their citation of the normal sciences literatures. Federal scientists cited range management titles twice as often as they cited Normal Sciences A and B.

Comparison of the citation rates for range management with the other groups of titles implies that following World War II, grassland ecology was split into two professional interests. The long-term

Table 9
Citation Rates of Range Management Literature
(Citations/In-print Year/Manpower Year)

Employer	1947–55
Academic	0.26
Federal	0.23

Clementsian synthesis of theory and practice was sundered, with practice split off, and range managers took most of the manpower out of grassland ecology. The publication rate of grassland ecology declined steadily after 1950 (see App. figs. 1, 2). The rates of citation, differentiated by employers, provide a preliminary explanation. The bulk of the specialty had shifted to technological concerns.

Conclusions

Comparing the bibliographical, sociological, and citation stages of grassland ecology from 1895 to 1955, we can perceive a generational rhythm to the growth and decay of the specialty. Originating in the 1890s in Nebraska, the specialty had by 1916 completed composition of its founding literature, mounted two competing microparadigm candidates, established two distinguished research and graduate training centers at Nebraska and Chicago, and generated regular use of its bibliography. Following World War I, the specialty entered a normal science stage of exponential growth of its bibliography, settled the question of which microparadigm candidate—the Clementsian—would guide the profession, incorporated federal scientists into the research efforts, expanded academic research centers throughout the Midwest, established a research front with an extensive coauthorship network, and by 1928 retired the founding literature. In the 1930s, in the midst of normal science research, the Great Drought generated an enormous test of the microparadigm. Citation patterns indicate that the specialty spun off as a separate group the federal researchers who had been incorporated in the 1920s, a split institutionally manifested in 1947 with the establishment of the Range Management Association. After 1949, the scientific specialty decayed, its rate of publication declined to zero, major scientists retired, and younger members cultivated economic problems of the grassland. This large cycle of scientific growth and decline was accompanied, of course, by intense intellectual development. The Clementsian microparadigm was established, but not without criticism. As we examine that criticism, we must, however, keep constantly in mind the picture of a self-referencing, inward-looking research community sketched by the bibliographical, sociological, and citation data. Only thereby can we appreciate how that community could withstand external attack, and discover its decay within.

6
A. G. TANSLEY:
A British Critic of Clementsian Theory, 1905—1935

The British Ally

We are not surprised to learn that Britain's leading Clementsian was a philosopher by temperament. Propelled by his father's fervent faith in education, Arthur George Tansley (1871—1955) was educated at the University College, London, and Cambridge University. Though he was always devoted to science, early being inspired toward botany, Tansley had a wide-ranging mind. Thus he designed his personal bookplates in 1893 to reflect his love of writers who rooted themselves in the English countryside: Wordsworth, Arnold, Meredith, and Lamb, and the botanists, Julius Sachs, de Bary, and Darwin, among others. As a young student at Cambridge University, he frequently strolled the countryside at midnight in conversation with his fellow student at Trinity, Bertrand Russell. They were friends through their early careers at Trinity, until Tansley's first job returned him to London. Tansley's student and friend, Harry Godwin, has quoted a colleague of Tansley at Oxford as characterizing him with the aphorism, " 'if you scratch a biologist you will find a philosopher.' "[1]

Philosophical inclinations led Tansley to a critical appraisal of Clements's ideas characterized by intellectual high-mindedness. He seldom chose to belabor Clements with empirical counterevidence, when he had the opportunity to strike to the principles of biological philosophy, to clarify conceptual terminology, or to straighten out illogic. His published reviews of Clements's books and his long, private correspondence with Clements were conducted with the best interests of ecology in mind, without narrow partisanship on behalf of his own ecological views. Spanning the entire life of the Clementsian theory and of grassland ecology, Tansley's critiques provide us with a convenient means of understanding the intellectual problems in Clements's theory which produced continual criticism and limited its acceptance outside the grasslands. In the end, even Tansley came to oppose the theory he had advocated. In 1935, he published a vigorous, nearly bitter, attack on Clementsian doctrine, "The Use and Abuse of Vegetational Concepts," which signaled the philosophical shift that was isolating Clementsian ecology. For many American ecologists, the Clementsian concepts of the biological community as a superorganism and the climatically determined monoclimax were not decisively rejected until the late 1940s and the 1950s, when experimental counterevidence was offered and anti-Clementsian theories revived. We shall see, however, how the Tansley critique combined with the long struggle with the 1930s drought to force grassland scientists to abandon central assumptions of the theory a decade before they were generally challenged by American scientists.

Tansley's professional debut came at a propitious moment in English botany. A tradition of amateur natural history was at its peak and needed professional scientific leadership. At the same time, the emerging specialty of plant geography required extensive, detailed, local inventories of plant species, a job that the local societies were situated to do. Tansley was a leader in harnessing the natural history effort of Britain at the national level for the purposes of professional botany. He used plant geography and later ecology as a cutting edge to change the profession.[2]

Disturbed by the lack of a "working" journal for British botanists, in 1902, Tansley financed and edited a new journal, *The New Phytologist*, as a forum for discussion and the presentation of work in progress. Inspired by the reading of Eugene Warming's *Oecologische Pflanzengeographie* and perhaps encouraged by a teacher's research into coastal vegetation, Tansley shifted his interests from plant anatomy to plant ecology.[3] The shift to the new field was completed by 1904, when Tansley called for the organization of a vegetation survey of the Isles at the British Association for the Advancement of Science. The British Vegetation Committee was established following Tansley's 1904 address, and significantly, in advance of the receipt of Clements's *Research Methods*. By 1905, when Clements's book was reviewed in England, Tansley had initiated the study of plant ecology in Britain and had established the institutions to conduct surveys. Clement's *Research Methods* arrived at the proper moment to provide a modern methodology and Tansley embraced it eagerly.

Clements met Tansley in 1911, at the International Phytogeographical Excursion in the British Isles, which Tansley organized to bring European and American ecologists together for an introduction to Britain's flora and for conferences over the common problems of their new specialty. They reinforced their friendship in 1913, when Tansley came to the United States for the second international excursion.[4] That same year, Tansley also led the organization of the British Ecological Society, of which he was the first president, in order to bolster the fortunes of ecology. A new journal, *The Journal of Ecology*, was published. Undoubtedly, the establishment of the British society spurred the establishment of the Ecological Society of America the following year.[5] In the years between 1904 and 1914, Tansley had successfully established the new specialty in Great Britain, justifiably earning the accolade from Clements in 1915, " 'You not only are the managing director, so to speak, of British ecology, but you are the outstanding figure . . . and thinker, which is much more important . . . in the whole European situation.' "[6]

As institutional innovator, Tansley appreciated and used Clements's *Research Methods*. The book met a need in British ecology created by the establishment of the vegetation survey. Tansley, and his co-reviewer, F. F. Blackman, hailed the book as "at once the most ambitious and most important general work on Ecology that has been published during the last seven years."[7] Their review of the book was significantly longer than any American review: two articles, taking twenty-seven pages. Clements provided the British ecologists with a methodology to undertake the survey they had previously committed themselves to do. At the Liverpool meeting of the Central Committee of the vegetation survey in November, 1905, Clements's methodology was officially approved by the committee and Tansley was given the task of condensing *Research Methods* into a pamphlet for the fieldworkers.[8] Several decades later, Tansley provided a book-length treatment of field methods, still fundamentally Clementsian, in *Practical Plant Ecology* (1923; 2d ed. 1926). The British ecologists clearly owed a great deal to Clements's *Research Methods*; a debt of this magnitude promoted tolerance for Clements's more risky philosophy of vegetation.

Tansley's orientation to ecology is difficult to characterize consistently, because his professional ambitions for the specialty of ecology led him to compromise on theoretical questions. We can see his fundamental ecological principles as derived from Warming and the mechanistic tradition. Ecology, he said, "is practically equivalent to the subject matter of Warming's book, a work that may be truly described by that often abused term, 'epoch-making.' "[9] Nevertheless, he modified this allegiance by accepting the reality of a vegetational social unit.

In the three decades after 1905, Tansley's acceptance cooled considerably, but nonetheless publicly remained stalwart until the sudden rejection in 1935. We can see the character of Tansley's commitment to Clementsian theory most clearly in his *Practical Plant Ecology*. This book came somewhat later than his adoption of the theory, because of several interruptions. Tansley was recruited for home ministry work (1914–18) during World War I, being too old for combat (he was forty-three in 1914). Following the war, he taught at Cambridge, but his attention had partially turned to psychoanalysis. He studied under Freud in Vienna and attempted to establish a practice in England, but, apparently, found dealing with patients' problems not entirely to his taste. Nevertheless, one result of this interlude was the publication in 1920 of a popularization of Freud's theories, *The New Psychology*.[10] Publication of *Practical Plant Ecology* in 1923, therefore, marked a return to the sort of concerns he had expressed in the vegetation survey before 1914.

Tansley announced his acceptance of Clements's main philosophical assumptions by describing the plant community, the basic unit of ecological investigation, as having its own emergent qualities. "Now, these plant communities have structures, activities, and laws of their own. Each has an internal economy depending on the relations of its individual members to one another; also an origin, history, and fate."[11] This statement might be considered innocuous enough, finding agreement by Clements and Warming. But Tansley was willing, however cautiously, to draw the parallel between the plant community and the human community that Warming explicitly had prohibited. "In these features [of plant communities] we recognize parallels with the nations, tribes, and societies of mankind, though the members of plant communities are not so closely knit as the members of human, and even of the high animal, communities, by a complex physical and psychical interdependence."[12] Tansley's reliance on the human society analogy had stood since his 1905 review of *Research Methods*. There, he had written that Clements's "view of vegetation as an organism is as legitimate as the familiar idea of a human society is from the same point of view."[13]

Tied to the idea that vegetational organisms had their own structures and followed their own laws was the notion that these structures were frequently not visible to unaided observation. This had been the fundamental insight of Pound and Clements in 1897 when trying to demarcate the boundaries of grass formations in Nebraska. Tansley accepted this thesis. The quadrat method was therefore necessary to discover what the eye could not see. In his 1905 review and in the 1923 book, Tansley practically repeated himself. Fresh with the new methods in 1905, Tansley wrote: "One of us has applied most of the methods in question to more than one type of formation . . . , with the result of acquiring the conviction that the very close attention to details of vegetation demanded leads to the recognition of features which would otherwise have escaped notice."[14] So in advising the novitiate to field study in *Practical Plant Ecology*, Tansley repeated: "But in other cases the interest of the distribution and the accuracy with which it follows the habitat conditions seem to demand an exact chart [quadrat chart], and the close observation required for this purpose may reveal the existence of other factors not at first suspected."[15]

One of the traditional hallmarks of the idealistic approach to biology from Goethe through Clements was the belief that the living organism could not be understood by study of an isolated stage or moment of its life, but only through a study of the entire progress of development. This message was the core of Goethe's *Metamorphosis of Plants* (1793), and Hegel's *Philosophy of Nature* (1816–27), and certainly was one key to Clements's theory. Tansley fully endorsed this point of view in a strong statement:

> It is the business of the student of vegetation to study these processes [competition and commensalism] and to trace out in detail exactly how they lead to the building up of the community. . . . It is well to start with a preliminary attempt to understand, in a general way, the structure and economy of the climax, the adult community, just as in studying a species or organism it is well to start from the adult form. But we cannot fully understand the significance of all the features of an adult except in the light of a knowledge of its development. The forces which go to the maintenance of the delicately adjusted equilibrium of a complex organism or a complex community cannot be estimated until we know how they come to be so adjusted, for they are largely masked by the adjustment itself. That is why the biologist

insists on the importance of the study of development, and the ecologist on the study of succession—the development of vegetation.[16]

The mechanistic biologist, in distinction to the organicist, did not need to study development before he studied the forces behind it. To the contrary, the forces could be inferred from analysis of habitat factors and the articulation of the forces from the direction of change. De Candolle and Darwin did not allow the teleology of a plant community's fate to creep into the requirements of understanding the community. They might have inquired impertinently of organicist biologists whether understanding of a plant community in the organicists' terms was even possible, for certainly the scientist could not directly study the entire development of a community. In Clements's vision, development on the Great Plains was initiated at the recession of the glacial sheets ten thousand years ago and had been concluded before he and his fellow grassland scientists were around to study the dominance of the grasses. Clements's system was in truth a vast web of inferences, dealing with the adult organism of the grassland in the rare places where it remained in its "natural" state and interpreting vast eras of paleontological remains.

It was interesting to note in his popularization of the "new psychology" of Freud and Jung, that Tansley provided the lengthiest explanation of the scientific necessity for studying the entire history of an organism. In the 1920 study, Tansley identified the "historical" approach to the organism as established by Darwin's evolutionary theory for psychology and sociology, as well as for biology. "Until we understand the history of the organism we are not really in a position to begin to estimate its existing significance at all, because we shall inevitably view it in a wrong perspective, misinterpret its characteristic features and misconceive its possibilities of future development."[17] He expressed two meanings of the phrase, "history of the organism," that had been intertwined in biology since the romantic biology of pre-Darwinian decades, the life of the individual organism and the evolution of its species, or "ontogeny" and "phylogeny" (which are terms that Tansley did not use). As we have already seen, the romantic dictum that the embryological development of the individual recapitulates the developmental history of the race was a central tenet of Clements's theory, which Clements stated in nearly its familiar form. Although Tansley explicitly derived the historical or genetic approach from Darwinian theory, the dictum antedated Darwin's *Origin* and was, in fact, used by Darwin for support for his theory. Evolution had not suppressed romanticism here.[18]

In applying Clements's theory of plant succession, Tansley had even a more formidable task than did Clements. Virtually no area of the British Isles was unaltered by its industrous residents.[19] "Fortunately," it was not necessary to have virginal nature to discover the laws of vegetational development. For Tansley, as for Clements, it was possible to perceive nature through the epiphenomenal effects of man: "It is true that ecological problems are complicated by man's activity—fresh factors have to be considered. But the plants themselves are working in the same way, tending towards the same effects, whether man is at work or not."[20] He reassured his lay readers: the teleological *tendencies* of "nature" remained unaltered by man, even if the appearances of nature did not. He had not quite slipped into a romantic organicism, but his popular language was close to it.[21] Darwin, in contrast, had stated forcefully in the *Origin* that there were only the appearances, the phenomena; there was no "reality" behind appearances, no "types" or "ideals" to which the messy "species" of scientific study conformed. But for Tansley, the beloved English countryside created by man's whim was a veil masking the "tending" of the vegetation itself, which was independent of mankind. There was not only the flower, but the force that drove the color through it. No doubt, Tansley would have paled before the characterization of his views as incipient romanticism or idealism, but I do not see any question that his effort to apply Clementsian ecological theory to Britain led to language and to conceptual formulations similar to those of overtly idealistic, twentieth-century scientists, Lloyd Morgan, Samuel Alexander, and Alfred North Whitehead. Not until the 1930s did Tansley move to scientific analogies and language that were not tinged with this vestigial romanticism.

A final indication in *Practical Plant Ecology* of Tansley's acceptance of Clements's fundamental assumptions was the subtle muting of Darwinian competition between allied species. Whereas de Candolle, Darwin, and Warming had referred repeatedly to the internecine warfare of abundant individuals for scarce resources, Tansley did not do so in this popular work. When he did, he usually moved quickly to discuss the developmental succession of the plant community.[22] This muting of competition for a popular audience deviated with his pronouncement in 1920, in the spirit of Warming, that competition was the fundamental mechanism of the plant association.[23] His perspective in *Practical Plant Ecology* was that of the relationship between the plant and its habitat factors. Thus, in his practical discussion of the peopling of new soils by plants, he did not mention competition.[24] Competition between species was called "decisive" only in a brief paragraph on the relation of plants to soil conditions.[25] The British plant world was not, for the amateur naturalist, red in tooth and claw. It was of greater moment that the plant community divided itself into strata, as between tall trees and shade-loving shrubs. In this paternalistic community, individuals of different strata did not compete and the high stratum aristocratically provided the conditions necessary for the lower to prosper.[26] If this was not the plant community as an organic society in analogy to the human community, then what was it?

If we have any hesitations in perceiving an organicism in *Practical Plant Ecology*, where Tansley adopted a popular and less strict language for his intended audience, these disappear when we turn to his professional reviews and papers, where he officially moved *halfway* toward organicism. Clements's overall accomplishment, from *Research Methods* of 1905 to *Plant Succession* of 1916, was so hugely useful to the science of ecology that Tansley was willing to move toward organicism on grounds of utility alone. Furthermore, Tansley was attracted by the sheer logical construction of Clements's *Plant Succession*. The concept of the formation-as-organism was central to the logical system in Clements; it permitted inferences about laws of organic development that other ecologists had not made. Tansley's student, Godwin, reminds us of his teacher's proclivities for philosophy: "There can be no doubt, however, that inherently Tansley was never the artisan concerned with laying the bricks and doing the joinery: his was the planning of the building, and his role from first to last that of the architect."[27] He quotes Tansley as characterizing himself as " 'by nature a dilettante with a strong interest in science and philosophy.' "[28] The foregoing characterization of Tansley with a philosophical bent relies heavily on Godwin's memoir for authenticity, but I had arrived at that characterization by reading Tansley's publications and correspondence with Clements before I read Godwin's paper.

Tansley praised *Plant Succession* for its logical and systematic qualities. "Professor Clements' large and important work is the first systematic monograph on the phenomena of succession . . . that has appeared, and if only for that reason marks an important stage in the progress of synecology."[29] Clements's major contribution was to provide a comprehensive theory. It provided, to use a term with which Tansley was unacquainted, a microparadigm for the new biological specialty. "There can be no doubt that the attempt to apply so clear and logical a system to actual vegetation will have a very beneficial effect."[30]

Clements opened both his preface and his first chapter of *Plant Succession* with unqualified statements of the formation as an organism, and Tansley did not complain. Clements repeated the assertion of *Research Methods*: "The developmental study of vegetation necessarily rests upon the assumption that the unit or climax formation is an organic entity."[31] If any reader might presume that Clements had perhaps wavered in his commitment to the idea even so slightly, since *Research Methods*, Clements left no doubt. After several years of testing his principles, Clements "felt that the earlier concept of the formation as a complex organism with a characteristic development and structure in harmony with a particular habitat is not only fully justified, but that it also represents the *only* complete and adequate view of vegetation."[32] Not only did Tansley not demur from this adamancy, but he defended it against its European critics. Continental botanists had reviewed Clements's conception of formational organism as "fantasies" "fairy tales," and—most unkindly—"laughable absurdities."[33] Tansley possessed a British temperament of compromise and hoped that the sparring botanists would rise above their one-sided and parochial views. Rational agreement on basic theory should be possible. Ecologists were particularly prone to interpret the world in terms of their backyards. Europeanists, studying the variegated Continental landscape with its many plant associations distinguished by soil differences rather than climatic differences, could not understand or accept an American theory based on the limitless view of the Great Plains, with its elemental dualism of grass and sky. Tansley believed that a rational and healthy personality involved a compromised balance of instincts and complexes, of biological drive and social training. Perhaps he believed that a rational and healthy theory in ecology involved a compromised balance of opposing views.

The Permanence of the British Landscape

The concept of the formational organism was less troublesome for Tansley in 1916 than Clements's insistence on the climatic dimensions of the formation. In *Research Methods*, Clements's concept of the formation referred to a small unit of vegetation delimited by habitat factors. Such a concept was tied to the physiological functions of the plant. In *Plant Succession*, the formation referred to the entire vegetative cover under a single, uniform climate, that is, to the climatic climax. In Britain, this meant that the fenland and the heath, quite dissimilar covers, were part of the same climax. In Clements's terminology, they were "subclimaxes," evolving at different rates, depending on local edaphic and artificial factors, toward the oak forest, which was the dominant plant association for the North Atlantic climatic climax. Tansley rejected the notion that British vegetation, as it then existed, could not be described as permanent (only the oak forest being permanent), and rejected also the idea that the major plant associations (e.g., forest, heath, moor, fen) could not practically be described as the "highest" unit of vegetation, that is, as formations. Logically, in terms of Clements's theory, restriction of the formation in Britain to refer only to the forest association might furnish "the most satisfactory scheme of classification," but leaving out "permanent communities of distinct life-form" would be, as a practical matter, an obstacle to the adoption of Clementsianism.[34] In 1916, the paths of succession were worked out only for a few plant communities, the North American midcontinental grassland, freshwater sand dunes, and the Rocky Mountain coniferous forest. Tansley was not ready to concede that all British plant communities would naturally evolve toward the forest. Consider the heath. Certainly heaths appeared to be maintained by grazing and cropping, but experimental evidence was not available to disprove the possibility that "subsurface" conditions, such as hardpan that prevented tree seedlings from establishing themselves, could also maintain the heath permanently and independent of man. A uniform climate for the British Isles could not be admitted to override diverse topographic and edaphic conditions. The lovely English countryside, the achievement of centuries of accumulated toil, intentional design, and gentle nature could not be merely a passing moment in the evolution of the oak forest.[35]

While accepting most fundamentals of Clementsianism, Tansley did have other criticisms in addition to the monoclimax. His paper, "The Classification of Vegetation and the Concept of Development" (1920) developed several of these. I consider this paper the most intelligent philosophical analysis of the new specialty published up to that time. He believed that the plant association, rather than the formation, as described by Clements, was the "natural" unit of vegetation. This assumption implied that, for Tansley, Clements's formation was a logical construction, rather than a natural entity; this implication was important in view of his idea that the formation was a quasi-organism, rather than a "real" organism, as we shall see in the next section. The plant association was a grouping of plants with uniformity in floristic composition, habitat, and physiognomy (appearance), and its definition had been standardized in 1910 at an international plant congress in Brussels. The formation was comprised of plant associations naturally found together. It is possible that Tansley's enthusiasm for logical system building had diminished since his 1916 review of *Plant Succession*. For in the 1920 publication, he declared that all classifications of plant life must begin with natural phenomena. "We are driven back to the vegetation itself, from which we ought to have started, as the only possible basis."[36]

He also usefully distinguished between succession and development. Clements considered succession to be the inexorable change of vegetation toward the climatic climax. Tansley preferred to distinguish between succession and development. Succession was the mere alternation of plant associations chronologically in the same area and was not necessarily progressive toward a climax, whereas development was the alternation of plant associations caused by the plants themselves when climate and soil remained uniform. Development was progressive, when it was not interfered with by agencies external to the plant associations. Cowles's classic studies of succession on the sand dunes of Lake Michigan were, by Tansley's definition, studies of development. Development was therefore a particular kind of succession, which he later called autogenic.[37] The distinction permitted him to define the English plant associations in heaths and fens, for instance, as permanent. They were the products of succession, influenced by habitat and edaphic factors, but they were not developmental, not progressing toward anything else, because the habitat "interfered" with the plant associations. By denying the monoclimax, denying the naturalness of the climatic "formation," and permitting nonprogressive vegetational change, Tansley was close to denying Clementsian theory. To do this, however, he had to attack the central concept of the formation as an organism. Three decades passed before he could make that ultimate denial.

Critique of the Formational Organism

Clements was privately shocked. Tansley made it precisely clear in a major publication of 1935 that he did not accept Clements's concept of the formational organism, either as objectively true or as a legitimate, heuristic device for doing ecology. Tansley apologized in advance in a letter for being "blunt." Clements admitted to his friend to being disconcerted: and why not? The title of Tansley's publication made a moralistic accusation: "The Use and *Abuse* of Vegetational Concepts and Terms."[38]

In retrospect, Tansley's thunder seemed long in coming. Since 1905, Clements had maintained with great vigor his thesis that the vegetative formation was an organism. Tansley had reviewed Clements's works, taking exception, it is true, to the concept of the organism, but never entirely rejecting it. He had analyzed the concept in a major publication, allowing himself a philosophical position that earned for him the label of "soft organicist." While he did not permit an ontological status for organicism, he had accepted it within science for its heuristic value. Tansley's motivations in waiting so long to undermine the philosophy of the Clementsian system were undoubtedly partly private, and so lost to our scrutiny, but in historical perspective, several changes were occurring in the 1920s and 1930s that made his blast understandable. First, as Tansley noted in his article of 1935, Clements's disciples and

Clements himself had been extraordinarily active in pushing their theory. The *Journal of Ecology* had recently published a three-part historical-philosophical statement by John Phillips, a South African ecologist, which went a fair way to making Clementsian theory ecological orthodoxy.[39] In one sense, if Tansley did not accept the fundamental assumption of organicism, then he had to speak again, quickly and strongly. Second, English philosophy had undergone a tremendous shift since the beginning of the century, away from technical idealism, which had indirectly offered support to the kind of position Clements espoused, to empirical realism. Despite a resurgence of idealism under the label of "organicism," in such philosophers as Samuel Alexander, Alfred North Whitehead, and Lloyd Morgan, empiricism had been given an important impetus by Einsteinian theory in the 1920s. Though empiricism in England antedated the impact of Einsteinian relativity, in the early work of Bertrand Russell, particularly, without question the new physics legitimated it. Tansley sensed the shifting winds of philosophy and desired to bring ecology within the bounds of current doctrine.

Pushing at Tansley's patience with Clements cum Phillips was the failure of international ecology to arrive, by the mid-1930s, at a uniform terminology or philosophical base. Controversy over Clementsianism was a perennial briarpatch that no one could eradicate. The International Botanical Congress of 1930 had created a commission on terminology to end the feud between Clementsian ecologists and European (Continental) plant sociologists, such as Braun-Blanquet, DuReitz, Ruebel, and others of the "Finnish-Swedish" school of phytogeography. To Tansley's consternation, Braun-Blanquet refused to join the commission, and the influence of his plant sociology was growing among younger European botanists. The internecine skirmishing seemed interminable; a single unified approach to the specialty appeared as far away as it did in 1900.[40]

A generation of controversy was enough for Tansley. He was facing retirement from the Sheradian Chair at Oxford in 1937, and still searching for the same logical system that he had called for thirty-three years earlier. He was trying to write his great work, the *Vegetation of the British Isles*, which required him to present a basic view on the vegetative formation. We can only surmise that these events and obligations pressed him for a resolution of the old problems of ecologists.[41]

Finally, Tansley's acceptance of the concept of the organism as a heuristic device for thinking about the formation always carried an implicit expiration date. By the 1930s the shift of philosophy from idealism to empirical realism had overrun Clements's views. Heuristic analogies were appropriate for a new specialty, when basic phenomena were only being recognized and the fundamental laws still unknown. Such crude analogies suggested hypotheses and clarified debate. A mature science should have no need for such devices. And if plant ecology was not mature in 1935, then it ought to be. Tansley's "Use and Abuse" article steamed with impatience. He implied repeatedly that some of Clements's thought was philosophically shallow and ought to be put behind the profession.

Let us trace the path of Tansley's criticism of the formational organism. Coauthoring the review of Clements's *Research Methods* with F. F. Blackman, Tansley raised his first objection to Clements's key concept in 1905. It was a logical confusion to consider the formation "real" in the same sense a table or a tree was real. At the same time, the scientist could legitimately postulate the concept of an "organism," if it aided investigation of the formation. The concept was acceptable for heuristic purposes. Clements's

> view of vegetation as an organism is as legitimate as the familiar idea of a human society from the same point of view. Both conceptions are useful and desirable so long as it is remembered that they are essentially analogical, that these quasi-organisms do not possess many of the essential features of real organisms.[42]

Tansley's rejection of an ontological status for the formation organism seemed straightforward. But he pulled back from asserting that the formation did not have "a real objective existence"; such an existence he believed it did indeed have. The question that Blackman and Tansley begged was this: if the formation had a real, objective existence, and the formation was not an organism, then of what did its real existence consist? The begged question was similar to another question that many biologists had begged since the publication of Darwin's *Origin of Species*—what was the ontological status of "species"? In the most nominalistic versions of Darwinism, "species" did not exist; what appeared to the taxonomist as "species" was only that: appearance. In the real world, independent of the biologist's classifying mind, only individual organisms existed; there were no "types" of organisms existing objectively in the same sense as individual organisms. Tansley rejected this nominalistic interpretation of the species question. He cited "the work of Jordan, of De Vries, and of the Mendelians" to indicate that there was a biological basis (in inheritance) for the reality of species *as a type*. While the causal mechanism for species maintenance in Mendelism was not known, he sanguinely expected it to be discovered. Similarly, though he did not know the causal mechanism behind formations, he believed that formations existed as something-like-organisms, and that their causal mechanisms also would be discovered. As he and Blackman concluded: "it follows that both species and formations have an objective existence, while all higher classificatory units are of the nature of abstract groupings."[43] Tansley gave no hint to what he believed the objective existence of formations consisted. For him, the universe apparently consisted of three kinds of things: entities (objects and organisms); minds; and—in this case—"quasi-organisms," which were not entities. It is a minimal reading of Tansley and Blackman, that they thought of the formation as an organism in that the plant could respond to the formation as well as to the habitat; to say less would be to say that the existence of formations made no difference. But even to say this little placed the coauthors closer to Clements's concept than they admitted.[44]

From this puzzling intellectual situation, Tansley took up the problem of the formation organism in his key publication, "The Classification of Vegetation" (1920). He defended Clements's concept of the formation as a natural unit, while again not agreeing that it was an organism. His defense was urgent, for Clements's approach had been attacked vigorously by the American botanist, H. A. Gleason, who argued that " 'the phenomena of vegetation depend completely upon the phenomena of the individual.' "[45] As I shall indicate below, Gleason provided the most trenchant nominalistic attack on Clementsianism ever, but, upon analysis, it turns out that he had simply shifted the problem that Tansley faced for formational quasi-organisms to "species." Tansley believed that the plant community was, as a unit, encountered and observed in nature as a naive experience, prior to the biologist's knowing any theory about the construction of vegetation. Plant communities "are certainly entities, in the sense that they behave in many respects as wholes, and therefore have to be studied as wholes."[46] As a unitary whole, a plant community had structure, organization, and functions, and communities of the same class responded similarly to similar habitats, even if the floristic mix might not be the same. Tansley clearly implied that a heath dominated by one shrub behaved the same as another heath dominated by a different shrub. The similar behavior was not a simple accumulation of the properties of the particular species comprising them.

Tansley further defended the analogy between plant and animal communities. It was permissible because it was a fecund source of provisionary hypotheses as to the structure and function of formations. Admittedly, the communities were different, and the difference stemmed from the animal organism's possession of psychic capabilities. This difference meant the ecologist had to be cautious in drawing the analogies. The human community was capable of active cooperation because its members were psychically aware of each other. The plant community was incapable of such cooperation and was, instead, integrated through competition and selection. And of course plant competition was conducted without awareness. A daffodil did not know it was crowding out a bluebell; the one plant simply grew and the other failed to grow in the sunny field.

Henry Gleason had raised similar criticisms of Clements's *Plant Succession* in his paper, "The Structure and Development of the Plant Association" (1917). But Gleason's solution to Clements's problems, unlike Tansley's, was in the spirit of Occam's dictum not unnecessarily to multiply entities in the philosophy of nature; he threw formational organisms out of ecology. Neither would he accept quasi-organisms. In botany, only the individual plant had ontological status. It was true that plant associations were encountered by the ecologists as units in nature, but this did not mean that they existed as beings. The plant association, or community, or formation, was simply the coincidental juxtaposition of individual plants in a selective environment.[47] "The association represents merely the coincidence of certain plant individuals and is not an organic entity of itself."[48]

In Gleason's universe, therefore, there were only individual organisms (and, presumably, physical objects). This position was philosophically untenable, as any nineteenth-century idealistic philosopher could quickly have shown, but Gleason was no more a professional philosopher than Clements or Tansley, and whistled his tune, oblivious to the cemetery of buried doctrines similar to his. As it turned out, Gleason did not quite mean that only organisms as individuals existed. In his more famous 1926 paper, "The Individual-istic Concept of the Plant Association," he amended his earlier position to mean that only organisms as individual species existed. Organisms behaved in a species-specific way: "every species of plant is a law unto itself."[49] This meant that classes of individual plants existed. Species certainly were to be defined in terms of genetic inheritance, but the mechanism of biological inheritance implied a relationship between individual plants (such as sexual reproduction) that could manifest itself as an individual plant behaving in a way typical of a class when interacting with an environment. What was the ontological status of this relationship between plants? Gleason did not say, but the problem was of the same variety as Tansley's problem of the ontological status of quasi-organisms. The reason, as idealists were arguing at this time, was that all individual organisms were sets of perceived relationships; they were never encountered except in their relations with other organisms. Clements's forma-tional organism and Tansley's quasi-organism were also sets of relations. The three men shared the ontological problem.

The development of genetic theory in the 1920s gave Tansley hope that a definition of quasi-organism could be obtained, in analogy to the definition of species in genetics. Here he had advanced over Gleason, who did not recognize the ontological problem with his concept of species. At the 1926 International Congress of Plant Sciences, he tried again to provide a conceptual basis for ecology upon which Clementsians and anti-Clementsians could unite. As his audience (which did not include Clements, who was unable to attend for personal reasons) well knew, he had "dealt with the subject elsewhere in some detail"; but he could not resist filing the latest evidence on behalf of the quasi-organism.[50] He compared the rela-tionship between the genetic mechanism of inheritance and the influence of environment in the ontogenic development of the indi-vidual organism to structure and development of the vegetative formation. In the 1920s, scientists were demonstrating that the structure of the growing organism was not determined solely by genes, but by environmental actualization of the genetic potential. Whether bones grew to their genetically potential length or eyes developed properly, for instance, depended on diets that included the newly discovered vitamins necessary for hormonal influence. Speaking only in general analogy, Tansley tried to find in the relationship between the formation and its environment a similar relationship as between genes and their environment. The "total stock of species" of plants in the plant community corresponded analogously to the genetic make-up of the individual organism. The analogy implied that the structure of seres in succession and of the climatic climax depended immediately on the constituent species, but the environment remained the ultimate determining force, pulling the development of this formation in one direction or another. Tansley thought that his audience "must all agree that ontogeny and developmental succession alike depend here and now upon the genes and species actually available, with their actual potentialities of response to the different factors of the environment."[51]

The argument significantly raised the level of sophistication of the plant and animal analogy from his earlier loose hypothesizing about similarities between plant and animal communities. Examining the new analogy critically, we must wonder what exactly he was driving at. How possibly could a scientist expect plant species in a formation to behave like genes in an animal? The chemical linkages of genetic inheritance were lacking in the plant community, for instance. It is likely that Tansley had no such detailed causal analogy in mind, since the great development in the biochemistry of genetic inheritance was more than a decade away. Rather, Tansley probably meant that a mathematical model or logical model of the formation, which permitted empirical inferences about the statistical character of the formation, could be developed, similar to the recently emerging mathematical model of genetics and natural selection, which permitted inference about the statistical distribution of characters in a species. "What we should not do is to treat the concepts so formed as if they represented entities which we could deal with as we should deal, for example, with persons, instead of being, as they are, mere thought-apparatuses of strictly limited, though of essential value."[52] While I have no evidence other than this paper that Tansley's mind in 1926 was heading in the direction of mathematical models for ecology, we shall see that a significant feature of his 1935 paper was precisely the proposal for mathematical modeling. Certainly, the level of expression in the 1926 paper does not allow us to say that Tansley had thought his way through the idea he was presenting, but he did seem to be working out of the dilemma of the ontological status of quasi-organisms. His formation as quasi-organism in 1926 was no longer quite a unit "of nature" or predicated to exist. It was a model that allowed the ecologist to look for certain effects in nature; it was the reality of the effects that should be of interest, not the "reality" of his model.

Despite Tansley's efforts in the 1920 and 1926 papers to ameliorate Clements's system and integrate it with emerging trends in biology, his inability to solve decisively the problem of the ontological status of the "organic entity" and the brevity of his remarks about the relation between concepts and reality worked against, rather than for, his purpose. Clements and his disciples were far more active in their elaboration of Clementsian doctrine than Tansley was in his criticism. And outright rejection of Clementsianism, as by Gleason, was not likely to persuade the Nebraska school. In 1934 and 1935, the South African grassland scientist, John Phillips, published a major, three-part review and defense of Clements's theory in Tansley's own *Journal of Ecology*.[53] Phillips's article was "the pure milk of the Clementsian word," but it was soured milk to Tansley's taste. His patience was exhausted. None of the warring factions in ecology were prepared to yield in their positions. In his astringent paper, "The Use and Abuse of Vegetational Concepts and Terms," Tansley attacked Phillips's article on nearly every point it made.

Not insignificantly, the volume in which Tansley's "Use and Abuse" article appeared was dedicated to Henry Chandler Cowles, who had recently retired from the University of Chicago. Tansley had been close to Cowles and admired the man as a scientist in a way, one may infer from tone, that he did not admire Clements. Cowles had been a follower of Warming, who had been also an inspiration for Tansley. When he needed examples of classic ecological work to illustrate his theses, as he did in 1926 for the Ithaca paper, he easily turned to Cowles's paper on the sand dunes of Michigan. He opened the "Use and Abuse" paper with a credit to Cowles as the first scientist to provide a "strikingly complete and beautiful successional series," whose work stimulated English scientists.[54] While this was true, it was a none too subtle way of saying that ecology would somehow have progressed without Clements's theory.

Tansley restated and elaborated all of the major points of his 1920 and 1926 papers. Phillips had sought to update the concept of the plant formation as an organism by describing the biotic community as a "complex organism" (a term that Clements had used in 1916, in addition to organism, to refer to the plant formation).[55] Tansley did not consider this progress; it was merely naming.[56]

The difficulty of Clements and his followers and of Tansley in resolving the problem of the ontological status of the formation-as-organism was not unique to them. It was shared widely among biologists, the philosophical foundations of whose science was in great flux in the first third of the twentieth century. The organic philosophers, Henri Bergson and C. Lloyd Morgan, for instance, revolted against mechanistic biology in favor of the view that living organisms had features that amounted to more than the accumulated properties of their atoms and molecules. Consciousness emerged in evolution as a novel property, not inferable from the molecular construction of organisms.[57] Morgan stated this view in somewhat contrived, though characteristic, language:

> at the upper level [of the evolutionary series] there seems to be a quite new kind of extrinsic relation—that which we speak of as cognitive— that which we regard as one of the distinguishing features of mind. The situation seems to be unique. Consciousness as supervenient is a late product of emergent evolution. But when it comes—at any rate, when it reaches the reflective level in us—we can contemplate what goes on at all lower levels.[58]

The organic philosophers' system involved an ontological shift of the level of biological reality from the molecular level organized in the cell to the level of the whole organism. As Clements had given the reality of the formation greater status than the reality of the individual plant, so the organic philosophers made the organism "supervenient" to the cell. In addition, their conceptualization of biological change bore similarity to Clements's concepts. During the course of evolution, qualitatively new types of being emerged, such as animals with supervenient awareness, yet the change leading to them was continuous, never qualitatively different. So also, in Pound's and Clements's description of 1898 of geographical change on the prairies: heading east to west, the species of grasses changed in the frequency of individuals imperceptibly and continuously, yet at some point the grassland changed from the tall grass formation to the buffalo grass formation.[59] For Clements, typological differences between formations, which had unique properties, frequently emerged out of continuous variation.

The concept of continuity was basic to Clements's thesis. He explicitly made the point I am making in comparing his views to those of Morgan: "the movement from initial stage to climax or subclimax is practically continuous."[60] The concept of continuity was also implicit in his choice of metaphors in *Plant Succession*. The formation "grows."[61] The quadrat reveals a "swing" of population.[62] Succession is a "series."[63] Succession can be traced in the "rise and fall" of each stage or sere.[64] Continuous changes accumulated to result in new qualitative stages; infinitesimal change led to a maximum, said Clements, choosing a mathematical metaphor.[65] "However faint their limits, real stages do exist as a consequence of the fact that each dominant or group of dominants holds its place and gives character to the habitat and community, until effectively replaced by the next dominant."[66]

Clements's critic, Henry Gleason, did not understand that Clements held concepts both of continuity of vegetation and of naturally limited organic boundaries. Gleason assumed that if vegetation were to be an organism, then it had to have a visibly discrete boundary, as—presumably—a plant or animal has an epidermal layer. Consequently, in his "Structure and Development of the Plant Association," Gleason asserted a continuity concept of vegetational distribution, according to which the plants in the boundary region (ecotone) of one association commingled with the plants of the neighboring association, in a density depending on habitat factors and without a definite "line" separating them. The "boundary" between plant associations was therefore more of a statistical transition area where the number of plants of one association decreased in a direction outward from the association's center and the number of plants of the neighboring association increased.[67] He did not understand that Clements accepted the continuum concept in analyzing the spatial arrangement of the ecotone and even the interior of the association. For Clements, the geographical transition between associations and formations (considered as organisms) was a continuum, and the chronological transition between seres in succession was also a continuum. What Clements could not accept was the notion that such a continuum prevented the appearance of a genuine community among plants of the association. He expected that at some point, the increasing density of plants in the continuum made a community possible. Clements's combination of continuum and type was typical, as we have seen, of a tradition in plant geography and was characteristic of twentieth-century organic philosophers. Organic entities were products of continuity in change. Clements would not accept the existence of continual gradation between plant associations as a decisive test of his theory, while he would have accepted the existence of community functions as such a test. It is therefore a mistake to contrast as absolutely antithetical the mental habits of thinking in terms of continuous change and in terms of typological classification.[68]

Tansley's Concept of the Ecological System

In "The Use and Abuse of Vegetational Concepts and Terms," Tansley's hesitant efforts to sketch an alternative approach to ecological philosophy finally produced his concept of the ecological "system." Tansley, in fact, coined the fashionable term, "ecosystem." Though he only sketched his views, he did state that his notion of "system" was derived from the philosophy of physics. It was clear, additionally, that he had ingested large helpings of the philosophy of mathematics. His ontological problem was now solved. The scientific mind did not apprehend "objects," or "organisms"—so he no longer had the problem of the status of "quasi-organisms." Rather, the human mind perceived only "events," or interrelated phenomena. The act of perception was primarily an act of "isolation," in which the mind seized a group of events out of the flowing stream of events with which it had fleeting contact. The abstracted "isolates" were "wholes," as were organisms, but they were not organisms. Galaxies, molecules, the ecosystem of a tangled bank, roaring trains, philosopher's tables: all were "isolates" in that some scientist was able, from the perspective of his specialty, to describe the systems they comprised as isolated from larger environments.

Systems were not artificial constructs of the perceiving mind. The elements in a system had to have functional integration and the system as a whole had to have stability; in other words, the inner relations between the elements of the system had to be sufficiently strong to withstand a certain amount of buffeting from changing systems or from the changing external relations of the isolated system. Scientists empirically and experimentally tested systems by pulling them apart to discover what relations were stable and what not. All systems overlapped; none were truly, physically isolated, in the sense of having no external causal relations (leaving aside the exception of the universe as a whole, for which all relations were internal).

From Tansley's new "systematic" point of view, systems—not organisms—underwent evolution. Some systems were more stable than others, in the sense that their internal relations were more capable of withstanding the strain of external pressures. Weaker systems broke up and strong systems persevered. "There is in fact a kind of natural selection of incipient systems, and those which can attain the most stable equilibrium survive the longest."[69] "Organic" evolution consisted of a "dynamic equilibrium," in which systems were organized (in the passive sense, referring to plants that did not have the power of self-initiation), persisted for some indefinite time, if they survived in the competition to be established, and finally broke up as external, centrifugal relations became too powerful. Their place would be taken by new systems. In the abstract, this scenario would be applicable to any aspect of the evolution of matter: stellar evolution, chemical evolution on the earth of the early forms of life, and ecosystems that were established by the earliest chemical relations of organic molecules.[70] The birth and survival of great wheeling stellar galaxies, and the familiar relation of farm cats and barn mice were both analyzable, in general verbal terms, within the same "system" approach. From this point of view, Clements's climax formation was simply the largest, most integrated plant ecosystem possible under the broad climatic regions of the earth.

In the brief sketch Tansley provided, the "system" approach would apply to an embarrassingly large number of theories. Echoes of Herbert Spencer's cosmic evolution and Whitehead's organicism reverberate through Tansley's formulation. But Tansley did not invite these echoes and he referred the reader to a popularization of science by H. Levy for a longer introduction to the concept of systems. Levy's account, *The Universe of Science* (1932; 2d ed. 1938), tied system theory to mathematics and physics. For Levy, the scientist was interested in systems that were "neutral" in their environment. This neutrality had two directions. In one, the system had to be neutral to the experimenter and the experimenter to the system; that is, the scientist's instrumentation or experimental design could not affect the system he was examining. "We do not choose a red-hot foot-rule to measure the length of a block of ice."[71] In the other direction, the system-plus-experimenter had to be neutral to the general environment. The ammeter used to measure the electric current in a wire had to be neutral to the circuit; both together as a system had to be neutral to the general environment.

Levy believed that no system was ever absolutely isolated from or neutral to the rest of the universe. The parts of the universe were connected. The "isolates" studied by the scientist were, therefore, partly artificial, partly natural. Nature was not a "seamless web," and where the scientist chose to tear the fabric was at a natural seam. The boundaries of the isolated system were determined by the question that the scientist was asking of nature. A biologist investigating the growth of a tree would isolate the tree as a system of growth at different boundaries than would a physicist attempting to determine at what pressure of wind a branch would snap off the tree. For the biologist, not only the soil, the air, and the other plant and animal life that interacted with the tree (in fact, the "ecosystem" of which the tree was a part while growing) would be the system in isolation. For the physicist, perhaps only the trunk, the branch, and the wind would be of interest, and the physicist might even be able to study the mechanical properties of breakage by sawing off a section of the tree and bringing it into a laboratory wind tunnel. In both cases, isolated systems effectively neutral to their environment were to be studied by the scientists.[72]

Absolute isolation was not a characteristic of systems for a variety of reasons besides the philosophical assumption that the universe was coherently connected. Levy pointed to the recent advances in quantum physics which indicated that at the atomic level, the experimenter must unavoidably interfere with the phenomenon he was seeking to investigate. "The very process of studying and measuring these elementary entities necessarily requires that they must be handled at such close quarters as to disturb the environment we normally presume to be neutral."[73] For purposes of macroscopic behavior systems (when we measure the height of a chair rather than the height of its constituent atoms), the interference phenomena of the atomic level could be ignored. But when we deal with atoms, their properties ("smallness, position, speed, and mass") are connected in such a way that the effort to measure one property changes the other. We could never find the atomic system neutral to the observer, but by specifying the interference of the observer we could neutralize the system comprised by the atom-plus-observer.[74]

The notion that nature was comprised of "systems" has a long and unclear history in Western science before Tansley's use of it. In the Newtonian tradition, the concept of system involved the idea that forces acting in opposition could produce relations of equilibrium between objects. Hence, the solar system was a stable balance between the gravitational forces toward the center of the sun and the centrifugal tendencies of the planets. A second line of thinking about systems has been slowly initiated in the twentieth century, incorporating the new idea of "feedback." The system as a whole has the ability to regulate processes to maintain equilibrium, as the human body, for instance, can regulate muscular and circulatory processes to maintain a constant internal temperature while the external temperature fluctuates. During the 1920s and 1930s theoretical biologists, such as A. J. Lotka and L. von Bertalanffy, attempted to make the concept of the self-equilibrating system the basis of a new biological theory. These efforts, however, went far beyond what Tansley wished to connote by his adoption of system theory for ecology. Bertalanffy had even gone so far as to insist on a new "organismic" biology, which was precisely what Tansley was trying to escape.[75]

What was the advantage of the concept of the system, as outlined by Tansley and Levy, for the specialty of ecology? Why should not Clements's concept of the organism have been as adequate, as long as ecologists knew what to do in the field? Tansley provided only one scientific criterion by which to choose systems over organisms; this was the criterion of scientific fruitfulness. For Tansley, the purpose of a concept was to provide the scientist with predictions or expectations of what he would find in nature. For this reason, Tansley had been willing to accept the concept of the quasi-organism. He wanted the concept of system for the same reason. He believed it would provide more verified predictions of behavior of plant communities than did the Clementsian scheme. Tansley rhetorically inquired of Phillips's concept of the complex organism, "what researches have been stimulated or assisted by the concept of 'the complex organism' *as such?"*[76] We may turn the question on Tansley. What uses did he expect to be made of the ecological "system"?

The answer is only implicit. Tansley wished to move ecology from dependence on a biological model to a physical model. This shift would have involved replacement of the growth metaphor of Clementsianism for the mechanical equilibrium metaphors of physics. He did not list the advantages to flow from this shift, but it is not difficult for us to conjecture what the list would have looked like. The first advantage would be, of course, an end to the ontological confusion that had traditionally befuddled biologists. This ontological clarification entailed, in Tansley's mind at least, clarification of description in biology. I think that Tansley was, in this, drawing upon Bertrand Russell's classic theory of descriptions and names.[77] The second advantage was that plant ecology would have been mathematicized in line with the new population theory of genetics. This unification of statistics and plant ecology would represent the second step in the mathematicization of plant ecology, a second step that Clements, who had taken the first step, refused to take. Third, ecological theory could be unified with contemporary advances in evolution. In all three cases, Tansley would have been pushing ecology in directions Clements explicitly opposed. Clements and his advocates, like Weaver, failed to hook their grassland paradigm to the rapid developments in genetics and evolution in the 1930s and 1940s. This failure was in part an ideological choice made by the grassland ecologists in the 1930s. It was a useful choice in the short run, but history placed scientific victories in the long run.

Clements, Conservative Biologist

System theory would have allowed the mathematicization of ecological theory. Plant ecology had been founded by the grassland school at one level of mathematicization, that is, the measure of the plant community and habitat factors. In the late 1920s and 1930s ecology was clearly ready for another step in mathematical advancement. This step would have to come through the concept of population, the numbers of individuals in a community, and would involve a thorough use of statistics. This step would have to come, that is, if ecology were to be synthesized with evolution theory, because evolution theory in the 1920s was rapidly progressing by using statistics. Ronald Fisher's classic work, *The Genetical Theory of Natural Selection*, was published 1929, six years before Tansley's "Use and Abuse" article. Comparing Fisher's work to Clements's *Plant Succession* and his publications of the 1920s, we glimpse how rapidly Clements's style, to say nothing of his metaphors, was becoming antiquated.[78]

Mathematicization was not easy. Clements opposed mathematical models in ecology, as was shown in an exchange with a paleoecologist, Frederick Frost. Frost prepared a report for the Carnegie Institution on statistical paleobotany, advocating statistical comparison of modern and fossil leaves. Clements blasted the report in a private letter, prompting a reply by Frost in which the beleaguered scientist took up Clements's charges quotation by quotation. Clements believed that statistics could only express known facts, but could not uncover anything new. Mathematics could only do, in other words, what he had pioneered at the end of the previous century: present precisely habitat factors for causal analysis. Clements opposed any notion of a mathematical model in plant ecology. Frost had proposed to pinpoint hereditary factors in leaves by comparing leaves for simple features (e.g., number of veins) of the allied species along the paleontological series. Leaves would be randomly sampled and constant factors (independent of habitats and geological strata, for instance) isolated. The set of propositions about what leaf features were hereditary and which developed by the environment would be a model, not a description of any one leaf or set of leaves, from which inferences could be made about the features of populations of leaves. Clements would have none of this. He believed that only experimentation could disclose which features were hereditary and which malleable by the habitat. It did not occur to Clements that Frost's statistical program was an experiment, that it expressed prediction and testing as much as Clements's work with seedlings in pots at Manitou Springs.[79] Clements confessed to "considerable enthusiasm" for "biometry," until it was found that the subject was inevitably more mathematical than biological, due to a reciprocal lack of knowledge of the other field." "It is difficult if not impossible, at present," he told Frost, "to base any method upon stable hereditary characters, since experiment can alone determine what these are."[80]

Clements's suspicion of Frost's approach also arose from his fundamental neolamarckian views toward evolution. He believed that plants and animals could acquire a wide variety and range of characteristics in their struggle to survive and adapt to their environment, and that these features were heritable. In the 1920s, he conducted experiments to transform plant species native to one ecological zone into a species adapted to another, higher, zone. Clements was quite convinced of the validity of his experiments, but this experimental Lamarckism fell to experimental disproof in the 1930s. It was this philosophical propensity that led Clements to stress the diversity of habitats to Frost, since this diversity, rather than genetic mutation, in his view, pulled species apart in the course of evolution.[81]

Neolamarckian opposition to Mendelism was in part political and social. Mendelism provided a scientific base for the eugenics movement and was allied to scientific racism. Clements opposed the eugenics movement, and of course did not agree with the fundamental assumption of the movement that heredity was intractable to environmental manipulation. The assumption contradicted Clements's liberalism, derived from his adherence to Lester Frank Ward's dynamic sociology. He laid out this point of view to his friend, John Phillips, at length in 1933:

> Apparently we have never discussed this topic, but as ecologists we must agree that environment is much more potent than heredity in the case of the individual and family. In fact, our experimental plants can frequently be made to "jump" a species or genus, and in a few cases an order. On the other hand, the power of "heredity" environment— such as our "little Italy" in New York or Chicago—is enormous, and it is this that is universally mistaken by the eugenist and layman for biological heredity. Johannsen was the clearest of all geneticists on this point, and he once said to me that no one was ever born a criminal. I have no serious objection to sterilization, for instance, but would much prefer to have it done as wholesale prophylaxis by means of ecologic control![82]

Political liberalism thereby ironically reinforced scientific conservatism. Tansley's situation, as we shall see, was the reverse: his political conservatism supported scientific innovation.

Tansley's attitudes toward mathematics in ecology underwent

change from the early part of the century to his "Use and Abuse" paper. Earlier, Tansley had deplored the strength of the mathematical description in English ecology and plant geography. He believed that it prevented English scientists from examining the long-term successional phenomena and discouraged them from theorizing about them. He was attracted to Clements's theory because it provided a means for ecologists to deal logically and even speculatively with their data, to raise them above descriptive mathematics.[83] By 1935, however, evolutionary biology was clearly going to be heavily statistical in methodology and conceptualization. The concept of system that he had embraced was easily treated in mathematical terms; in fact, it had so been treated for three hundred years in classical mechanics.

Recommendation of Levy's *Universe of Science* signaled Tansley's desire to incorporate more mathematics into ecology. For Levy was a mathematician and his book emphasized mathematics as "the queen of the sciences," explaining how an apparently nonempirical discipline could be so fruitful in any empirical science. Mathematics provided the means of symbolizing the system "isolates" studied by scientists, whether of measurable quantities of physics such as velocity of a falling body or of biology such as the distribution of genetic traits in an interbreeding population. The mathematician developed a "general case" as a deduction from a mathematical generalization, which the scientist could test against a specific empirical case. The law of falling bodies can be deduced from the so-called "times-squared" and "odd-numbers" rules and then verified in experience by rolling balls down inclines. In one version of the history of science, of course, this is what Galileo is reputed to have done. In the biological sciences, similarly, from mathematical definitions about reproduction and food supply, deductions could be drawn about the size of a stable population (Levy did not provide biological examples).[84]

The point of all this was to integrate ecology more closely with the theory of evolution. Ecology had been founded on the assumption that plants as vegetational coverings evolved from less complex to more complex forms. Warming, Tansley, and Clements agreed that the evolution of communities depended on the struggle for survival of the plants, though they differed on the question at what points the struggle was crucial and to what extent to emphasize the struggle. Clements, no less than Warming, perceived evolution as the great movement of plants. But the development of plant ecology in the hands of the Clementsians had become isolated from the general development of evolutionary theory. Clements was not going to utilize a population genetics approach. Nevertheless, ecology contained concepts that were to be useful in synthesizing ecology with modern evolutionary theory, such as "habitat" and "niche." Indeed, in animal ecology especially, speciation remained a viable problem, though in plant ecology it was not of central concern. Tansley quickly picked up the work of Charles Elton, who was trained under Julian Huxley in the 1920s at Oxford University, where Tansley took the Sheradian Chair in 1927. Elton's pioneering *Animal Ecology* (1927) developed the concept of population ecology for animals and Tansley saw the obvious importance for plant ecology. In a review, he wrote:

> The problem is to understand in detail in particular cases how the spread of the new genotypic variations is effected. Mr. Elton suggests that the great fluctuations in animal numbers may provide a partial key to the solution. If new heritable variations arise in a population whose numbers are at a low ebb, so that the struggle for existence is partly suspended, then, when the great increase in numbers arrives, the new variation will automatically spread very quickly, and a very large proportion of the new population will possess it. It is clear that ecological work alone can settle such questions, and is therefore essential as a contribution to any complete understanding of the mechanism of evolution.[85]

Since vegetation was the crucial framework within which the animal food chain operated, the great historical movements of vegetation, which the Clementsian scientists had been elaborating, provided necessary knowledge for understanding speciation.

The outstanding experimental work of the 1920s on population of species and habitat variation as an isolating mechanism was being conducted by Theodosius Dobzhansky with fruit flies—*Drosophila*. Clements closed out the possibility of synthesizing Clementsian ecology with similar studies in plant ecology by explicitly rejecting, in his personal correspondence, *Drosophila* studies and all single-species studies as adequate bases on which to erect an evolutionary theory.[86]

The Political Ideology of Plant Ecology

Not only the pure milk of Clementsian doctrine, when poured from Phillips's pitcher, soured in Tansley's mouth; Tansley tasted in the Phillips articles an unpalatable political flavor. He was moved to conjecture:

> It is difficult to resist the impression that Professor Phillips' enthusiastic advocacy of holism is not wholly derived from an objective contemplation of the facts of nature, but is at least partly motivated by an imagined future "whole" to be realized in an ideal human society whose reflected glamour falls on less exalted wholes, illuminating with a false light the image of the "complex organism."[87]

Printed in a professional scientific journal of ecological research, detached from the social and political concerns of the era, Tansley's jibe reads like an abstract pique directed at university politics. Of course, nothing could be further from the truth. We need only recall the state of European and British politics in 1935 to understand the possible references in Tansley's criticism. In 1935, Europe was tearing itself between fascism and communism, with the bourgeois liberal states, France and Britain, anxiously looking on. Premonitions of a second world war between the ideologically opposed states were frequently stated, and it had been two years since the Oxford Union moved its famous and infamous pacificist declaration. The British left was nearly at its peak of popularity in 1935, with disaffection with the Soviet Union still in the future. The national British government was in reconstruction; Ramsay MacDonald, prime minister since 1929, resigned in June, 1935, and elections (which held onto a conservative majority) were held in November. English intellectual and political life was politically charged, and the major political questions were economic and ideological.

The popularity of socialism, in some form or other, brought "organistic" social theory to the front of ideological discussion, in opposition to the traditional individualistic and competitive ethic. Leftist intellectuals argued that society should be viewed from the perspective of the people as a whole, rather than from the egocentric perspective of the competitive, privatistic, and individualistic bourgeoisie. We need only to recite the names of intellectuals and artists of the 1930s who identified themselves with socialism to be reminded of the potency of the organistic approach in politics: Auden, Spender, Day Lewis, Isherwood, among poets; Arthur Koestler, H. G. Wells, George Orwell among novelists and critics; J. B. S. Haldane among biologists.[88]

What does this tell us about the political attitudes of Tansley? Little, directly; but it does indicate that he opposed believing in scientific propositions for political reasons or out of political hopes. We can safely infer Tansley's opposition to political holism and the notion of an organic society (in contrast to a liberal society). We can infer nothing about Tansley's specific political beliefs, whether he supported the conservative national government of MacDonald, for instance. But it is not necessary to know on what political platform Tansley stood. The general tenor of his political attitudes can be pieced together from correspondence, publications, and activities. We would of course undoubtedly benefit by having an extensive collection of his personal manuscripts.

The fragmentary memoirs of Tansley speak of a man individualistic by temperament and by intellect. He undoubtedly believed that human society was permanently stratified by class and race, and that these strata could not be ameliorated without loss of individual identity. We know this, because Tansley defended the notion of ecological stratification in forests, for instance, by reference to it in human society.[89] In his memoir on Tansley, Harry Godwin sketched a picture of his teacher, one evening in 1929, loose with Chianti wine imbibed at an Italian restaurant, "philosophizing before an opulent florist's window that he supposed with the progress of social reform we might expect such shops to disappear and he added, 'and I, for one, shall regret it.' "[90] One hesitates to make too much of this anecdote, but certainly it implies a commitment to the individuality of Britain's shopkeeper culture, as well as a fondness for innocuous loveliness.

More to the point was Tansley's lifelong admiration and support of Freudian psychology. Freud's perspective on human personality formation and behavior was anathema to most Marxists during the interwar period. During the 1930s in Britain, the personal struggle of English intellectuals, almost all of whom were middle class in origin, was to free themselves from the "bourgeois" habit of psychologizing human behavior. Rather, behavior was to be explained in Marxist terms of structural dynamics of society. Historians today frequently see Freudian psychology as the epitome of bourgeois analysis of bourgeois neuroses. Tansley had to have been aware that his, or anyone's, acceptance of Freudianism separated him from a major part of left-wing analysis of English society in the 1920s and 1930s.[91]

Tansley approached the scientific institutions of his era as a reformer within the establishment, not a rebel outside. He did not wish to destroy the English scientific and university establishment, only to energize and motivate it in the direction of modern scientific theory. In the 1910s, he attempted to reform the teaching of botany, to incorporate an ecological approach as basic to understanding plant life. This curricular endeavor awoke fierce opposition from academic traditionalists. Clements, as part of the Bessey avant garde, had experienced the same fight when he participated in a similar academic reform a decade earlier. He had sympathy for Tansley. His establishment of the Vegetation Committee was an effort to harness local botanical efforts in a national survey, to raise theory by raising the level of research. In 1941, he was one of the three founders, along with John R. Baker, a fellow biologist at Oxford University, and Michael Polanyi, chemist and antipositivistic philosopher at the University of Manchester, of the Society for Freedom in Science. The society opposed state planning of scientific research for utilitarian purposes, opposed scientific planning of society by socialists, and defended the bourgeois society that paid for their researches and left them in intellectual freedom.[92]

Tansley's involvement with nature conservation similarly reflected the views of a man within the establishment. The British Nature Conservancy Movement originated in the 1940s, but had active roots two generations earlier. The motivations for preserving the English "wild" were partly scientific: Tansley wanted to save sufficient amounts of nature from the expanding urban and industrial centers for scientific research. As he stated in his 1944 plea for preserves, "The maintenance of adequate samples of wild plant and animal life is necessary to ecological work, for they furnish its basic material."[93] It was a similar motivation that led scientists in the United States to support the establishment of Yellowstone National Park and the other natural preserves, and brought scientists in Europe at the turn of the century to organize themselves as pressure groups on behalf of nature preservation.[94] Tansley's *New Phytologist* and *Journal of Ecology* regularly covered the preservation movement's activities in Europe and in the United States.

The English landscape was more than the object of scientific study. It also accounted for a considerable part of the uniqueness of the English character and culture. Tansley believed, as did so many of his countrymen, that the countryside had to be preserved to preserve the national identity.

> This combination of cultivation with half-wild and wild country is one of the most precious parts of our national heritage, for nowhere within so small a space as the area of Britain is there a greater variety of rural beauty. And it is with this background—gradually changing as the centuries have passed—that our history has unfolded and our culture developed.[95]

World War II, which had destroyed so much of the British cities, ironically forced the British to destroy large plots of their countryside as well. Forests were cleared, fields ploughed, urban factories dispersed to the country, and military camps and air fields disrupted the pastoral scene. After the war, Tansley was involved in planning new towns that would consume still more of the limited rural space. "How much of [our] unique inheritance can we preserve in the years of profound change that lie ahead of us? Not all of it, certainly."[96] He did not say, "But we must preserve as much as we can." He did not need to.

Tansley led the nature conservancy as one of the British scientific elite. He was a member of the Royal Society of London. He was active in the movement that led in 1949 to the chartering of the nature conservancy as an advisory group to the Privy Council. He was its first chairman (1949–53).[97] He sought to change those aspects of English society and life with which he was concerned by voluntaristic methods and by utilizing his expertise as an advisor to the government. Nothing in what I have read of his writings or in his correspondence with Clements indicates any propensity for leftist politics or for a left-wing critique of British society. His prickly jab at Phillips's political holism derived from a simple antipathy to an ideological persuasion that he did not share. He did not want to see ecology as a scientific discipline justify leftist political theory. He referred to Phillips's social holism with barely concealed sarcasm.[98]

Regarding the ideas of the leading exponent of holism, J. C. Smuts, Tansley thought "a good deal of fuss is being made about very little."[99] While he was willing to concede some usefulness philosophically to Smuts's ideas, he would not yield on the question of their value to science. Smuts's theory seems to have bothered him considerably; he referred to Smuts by name three times in the "Use and Abuse" paper within the context of criticism of Smuts's holism. Smuts was a South African national leader, who wrote his book, *Holism and Evolution*, after being retired from politics. By "holism," Smuts meant that there was a creative tendency in the universe for matter and for life forms to aggregate themselves through evolution into larger, more complex, and more integrated unions. Atoms cohered into molecules, molecules into cells, cells into organisms, organisms into communities.[100] When his book was published (1926), Smuts was immediately put into the camp of English idealists, Samuel Alexander, Bradley, Whitehead, and J. B. S. Haldane (who loudly praised the book). Though the book did not address problems of politics and value choice, Smuts intimated that these problems could be approached from the holistic point of view. He described, albeit vaguely, holism as the source of all values. He believed that the highest end of personality development was the expression of the whole personality or whole soul, contrasting that to the attainment of simple hedonistic pleasure or avoidance of pain. He implied that political union of peoples above the level of the nation state was better than nationalism, and perceived the League of Nations as a first step in that direction.[101]

The generality and vagueness of Smuts's notions might lead us to see in them only harmless good intentions for the general betterment and moral improvement of men. A conservative political critic, like Karl Popper, however, saw in Smuts an echo of Hegelian authoritarianism. He labeled the views of Smuts, Alexander, Whitehead, and others as the "new tribalism" which sought to impose a "higher idealism" and "higher morality" on the individual.[102] If Popper could read authoritarianism in Smuts, no doubt Tansley could have objected—though we do not know this directly—to the implicit antiindividualism. Tansley was undoubtedly familiar with Smuts's book before he was provoked by Phillips's articles on Clementsian doctrine to attack it. Smuts lectured at Oxford, where Tansley was teaching, in 1929. I cannot prove that Tansley did in fact interpret

Smuts's book in terms of totalitarianism. Smuts did not directly deal in *Holism and Evolution* with political theory, but in Europe in the 1930s, lines were being drawn between opposing ideologies and the benefit of the doubt was frequently left out of discussion. Tansley intimated that Phillips's ecology was in the political service of socialism and, there is reason to believe, given the charged atmosphere of this declining decade, that Tanley rejected holism and an ecology with holistic concepts as politically and socially unacceptable.

7
SAVING THE PRAIRIES:
The Great Drought of the 1930s and the Climax of the Grassland School

The Roots of John Weaver's Faith

The Great Drought devastated midwestern agriculture in the worst year of the Great Depression. In the summer of 1934, the level of available water on the plains had fallen so low that no plant could reach it. Without cooling moisture, soil temperatures frequently rose above air temperatures, and in that summer, surface soil temperature "not infrequently," as Fred Albertson calmly put it, reached 140°F. Air humidity fell 15 percent to 25 percent: air shockingly dry to midwesterners used to the humid and thunderstorming prairies of summers past.[1] Many regions of the midcontinental grasslands lost from 50 percent to 95 percent of their basal plant cover by 1935.[2]

Farmers were forced by economics into the worst possible response to the drought. After 1930, the price of wheat dropped precipitously from the high levels of World War I and the 1920s. Farmers compensated by shifting their capital investment into cattle raising. The need for pasturage led to overgrazing precisely at the moment the grasslands were weakened by the drought. Consequently, it was pasturage that the drought first destroyed. The short-rooted grasses preferred by livestock died in dry winds and dust burial. Vast counties of the high Great Plains in western Nebraska, Kansas, Oklahoma, and eastern Colorado were abandoned.[3]

From 1933, when the drought began, until 1942, when rains providentially returned in time to produce agricultural abundance for the American war effort, the grassland school of scientists was preoccupied with the drought crisis. Scientific research was intently focused on the twin problems of range management and grassland conservation. An extraordinary peak of scientific progress was reached within the Clementsian microparadigm. Preoccupation with the drought also, ironically, isolated the grassland school from the new currents of ecological theory and drained its scientists of confidence in the ultimate recovery of the original prairies.

The leader of the academic research effort in the struggle to save the Great Plains was John E. Weaver, Clements's student and collaborator at the University of Nebraska. Weaver trained more of the academic scientists engaged in the drought than any other individual. His collaborative research with Clements and his own research provided a substantial portion of the disciplinary knowledge available to meet the crisis. He undoubtedly influenced many hundreds of young scientists whom he did not know through his widely adopted textbook in plant ecology, coauthored with Clements. It was therefore of consequence that this distinguished leader had his faith in the Clementsian microparadigm shaken by the struggle with the drought. The Clementsian microparadigm proudly withstood the assaults of critics in England and Europe, but it could not hold up when challenged from within.

Weaver's commitment to the Clementsian microparadigm was long standing, undoubtedly originating from his undergraduate days at the University of Nebraska and deepened in his doctoral study under Clements at Minnesota. The depth of commitment was not only determined by the intellectual attractiveness of Clements's analogies, but also by Weaver's long experience in field research on the plains. His professional debut was made with pioneering research on root systems that validated Clements's theory, and which required arduous hours of physical labor examining grassland plants. After a brief teaching stint in Washington, where he had undertaken research on the local prairies, Weaver returned to the University of Nebraska and initiated a half-decade's study of the root systems of the grasses, forbs, and shrubs that inhabited his Nebraska grassland. These studies convinced him that grassland plants behaved as a community, not as individuals, and that the entire midcontinental grassland represented one vast formation.

Just as Pound and Clements had discovered in 1897 that they had to count every individual prairie plant to understand the transition from one grass association to another, so Weaver had to excavate individual plant roots like an archaeologist, physically separating out their taps and branches, measuring and mapping the systems, to understand how root systems knit together and responded to climate. In 1917 and 1918, Weaver dug more than 100 pits, three feet wide, six to ten feet long, and six feet deep. Root systems were then segregated with a hand pick. He excavated about 1,150 individual plants, representing about 140 species, in eight geographically separated plant communities.[4]

There in the trenches Weaver learned the reason for the great stability of the grasslands. Even abnormal weather for two years was not "likely" to change fundamentally the vegetation, so great was root control of the soil and ability to adapt to moisture content.

> Vegetation is not only an expression of present conditions, but also to a greater extent a record of conditions that have obtained during a period of years, and the record is not likely to be altered greatly in a year or two in which conditions may depart from the normal . . . . Many prairie plants absorb moisture well beyond a depth of 5 feet, while soil-moisture extends many feet beyond the greatest root depth.[5]

Many species had the capacity to adapt their roots to moisture conditions. In soil in which available water extended deeply, roots strove downward with little lateral branching; in soil in which water was usually available only on the surface, lateral development was pronounced. Writing of *Chrysopsis villosa* (Hairy Golden Aster), he illustrated this adaptability:

> This affords a very clear case of the effect of the habitat upon root development. The plains form has a root which is approximately twice as deep-seated as those in the sandhills. While the tap is prominent throughout in the former, it soon loses its dominance in the sandhill form and often scarcely exceeds in importance some of the stronger laterals. While both forms are supplied with rather abundant surface laterals, in the plains form these are largely confined to the surface root, while in the sandhills they occur to a much greater depth and are abundant along all of the major branches.[6]

The capacity of many grassland species to adapt their root systems to moisture conditions permitted the community to adapt to the climatic environment. "Each community, viewed as a whole, has its own root habit, the one best fitted to the particular environment."[7] The tall grass prairie of eastern Nebraska had deep roots, while buffalo grass plains to the west had extensive lateral rooting in order to take advantage of the frequent, brief, summer showers.[8]

Pound and Clements early believed that the prairies and plains grass communities were distinct formations, that is, different climax plant covers. Weaver's doctoral thesis contributed evidence that both communities were but parts of a single formation. In his study of the prairies of southeastern Washington, Weaver demonstrated that its *Agropyrum* consocation was related to the *Stipa-Agropyrum* association of the plains.[9] Clements and Weaver became convinced that a single set of plants dominated all grassland communities. Student and teacher tested this thesis for more than four years (1920–23) at stations located in three states to represent true prairie, mixed prairie, and plains climatic associations and at a variety of edaphic stations. The purpose of the experiments was to determine whether the grassland dominants could cover the climatic range necessary to enable one set of dominants to spread over all grassland associations. The results of these experiments were reported in Clements's and Weaver's coauthored *Experimental Vegetation, The Relation of Climaxes to Climates* (1924). This work provided the most thorough application of Clementsian theory to any plant formation.

The experiments on the range of the grassland dominants showed Clements and Weaver that the true prairie and the buffalo grass plains were only subclimaxes, not terminal climatic formations. Following the recession of the last glaciers, the original grassland mass was rather uniform, somewhat resembling the contemporary mixed prairie (both tall and short grasses) that existed between the prairie and plains communities. The short grass plains had resulted only recently from overgrazing of this original mixed prairie; tall grasses were cropped, leaving the short grasses to dominate the landscape in semiarid areas where their rooting system was favorable.[10] The implication was that in the natural course of grassland succession, without man's intervention, the true prairie would extend its domain back over much of the territory dominated by the short grasses.

While there were many interesting scientific issues in *Experimental Vegetation's* support of Clementsianism, we want to notice mainly how Weaver (along with Clements, of course) was confirmed in his belief in the stability of the grasslands. He could not conceive of circumstances under which the plants might naturally be destroyed, or under which maintenance of the plains structure would require human intervention (rather than withdrawal). Indeed, in 1922, he could not conceive of a natural catastrophe like the ten-year drought of the following decade. His imagination, like anyone's, was largely limited to versions of his own experience. He labeled an abnormally dry fall and winter (of 1922), a "great drought."[11] At the true prairie station outside Lincoln, 22 percent of spring seedlings were killed by the drought.[12] He could not imagine the circumstances a dozen years later, when drought would kill up to 95 percent of established cover.

In 1934, after two years of drought, Weaver published a paper that summarized what he and the grassland school had learned about the American prairie in the preceding decades. The paper provided a good opportunity to change his views on the stability of the prairie, should the current drought have led him to modify them. Weaver, however, remained confident. He described how prairie that had been disturbed to a depth of six inches could return to its original composition "within 25 to 30 years."[13] The original prairie, although scarce, was still to be found in small isolated acreages. Yet, these spots were disappearing. He did not feel, however, that ultimate disappearance of the prairie could occur, with replacement by some invading plant (such as Kentucky bluegrass). The prairie was too tough and stable for that. Temporary disturbances by man could be overcome. "Sometimes [after] prairie land is broken, the land 'goes back' to prairie," he wrote in another article of that year.[14] His optimism shone through the qualifications of his generalizations: "Unless disturbed by man, and barring the entrance of a new dominant from another region, the prairie will maintain possession until there is a fundamental change in climate or a new flora develops as the outcome of long-continued evolution."[15]

Only one premonition of disaster was expressed by Weaver in 1934. In his paper on the "Stability of the Climax Prairie," he cited a paper by another scientist—not drawing on his own experience, interestingly enough—on the problems incurred by upsetting nature. "[Smith] has recently pointed out some penalties for upsetting the balance of nature in the prairies and plains . . . . He concludes that man, the disturber, will have to employ artificial control efforts for a long period, or be seriously handicapped in his labors."[16] Smith's thesis, that wild nature can exist only if actively conserved by man, had been learned in the previous decade by some other scientists, especially in forestry, but it was a lesson that the grassland scientists would need more years of drought to learn.

The Faith Erodes

In what ways did Weaver's long struggle against the great drought of 1933–42 change his views on the Clementsian philosophy? He had studied drought before, but the localized and seasonal drought of 1922, for instance, did not so alter the structure of the prairies that he had to alter his view of them.[17] The severity of the Great Drought was much greater and of longer duration and the destruction of the plains consequently greater. Weaver had the opportunity to reconsider longheld views when he and Clements revised their classic textbook, *Plant Ecology*, for a second edition published in 1938. Comparing the first edition of the text (1929) with the revised edition, several important thematic changes are noted. First, Weaver and Clements removed references to the vegetational formation as an organism. Second, they admitted that man could *permanently* alter the structure of the grasslands. And, third, they consequently qualified their axiom that all succession was inevitably progressive toward the climatic climax. These changes concerned the basic tenets of Clementsian philosophy. Whether Weaver retained a privately held commitment to the axioms of his teacher's system from which he publically retreated, I cannot determine without manuscript evidence. I believe, however, that the language of the publications of the late 1930s and afterwards were of a scientist who had resigned himself, in the middle age of his career, to the error of the fundamental postulates that had once guided his intellectual life.

In the first edition of the textbook, Weaver and Clements strongly stated the axiom that the formation was an organism. [1] "Each formation is a complex and definite organic entity with a characteristic development and structure."[18] The formation was an individualized organism, just as was the individual tree. [2] "Just as a tree passes through the seedling and sapling stages and then grows to maturity, so, too, the formation arises, grows, matures, and dies. The formation, moreover, like the plant, is able to reproduce itself, as may be seen after fire, lumbering, or other catastrophe to the vegetation."[19] These refrains were familiar in the Clementsian litany, but we must remark that in 1929 they were unchanged after criticism from Tansley and from American critics like Gleason. In the second edition, however, the second statement was dropped entirely. The formation was not an organism just like a tree. Abandoning the tree analogy was more than simply a literary change; it was an abandonment of the ontogenetic analogy that had underlain Clementsian philosophy since *The Structure and Development of Vegetation* (1904). This analogy had been, as well, at the base of Tansley's acceptance of Clementsianism. Tansley did not, we recall, believe that the formation actually was an individual organism, but he incorporated the analogy of formation to organism for the provisional hypotheses it suggested about the way formations worked. Abandonment of the ontogenetic analogy, therefore, was tantamount to abandoning the search for laws of vegetative development. The analogy suggested that succession must necessarily be progressive, just as growth must be progressive; without it, the necessity of progressive development disappeared.

The loss of necessarily progressive succession in Clementsian philosophy immediately showed itself in the admission by Weaver and Clements that man could *permanently* halt vegetative development toward the climatic climax. This admission put them in agreement with Tansley's and other critics' position of previous decades, that it made no sense to call the climatic climax the only truly permanent climax, when in fact much of the landscape was permanently fixed by interaction with man in a state that was not a climatic climax. Tansley rightly looked at the British landscape in this way. Weaver and Clements phrased the matter in the following passage from the 1938 edition, which was not in the 1929 edition:

> While the climax is permanent because of its entire harmony with a stable habitat, the equilibrium is a dynamic one and not static. Superficial adjustments occur with the season, year, or cycle. . . . But these modifications are merely recurrent or indeed only apparent. While change is constantly and universally at work, in the absence of civilized man this is within the fabric of the climax and not destructive of it. Man alone can destroy the stability of the climax during the long period of control by its climate. He accomplishes this by fragments in consequence of destruction that is selective, partial or complete, and continually renewed.[20]

They expanded their discussion of man's influence on the environment to include an indirect reference to the drought. Men removed the natural cover in clearing, lumbering, and cropping. They substituted a crop cover for the natural cover. And they regenerated the natural cover as a self-conscious act of reconstruction, or imposed a new natural cover. Removal and agriculture were necessarily destructive. They opened the soil to wind and water erosion. No doubt, Weaver had in his mind the terrifying picture of dust storms on the prairie.[21]

Weaver and Clements still maintained that vegetational succession, when independent of man's interference, was progressive, but in the second edition of *Plant Ecology*, it was difficult to find nature independent. Indeed, a statement that succession was progressive was unwittingly illustrated by an example of the artificial invasion of Russian thistle, which was introduced to the Dakotas in agricultural commerce. The illustration implicitly contradicted the point it was supposed to illustrate.[22]

If destructive "coaction" by man permanently halted vegetational development toward the climatic climax, or destroyed the climax, then the obvious inference was that only man could put the succession back in the direction of climatic development. Man so inextricably wove himself into the ecological fabric of nature that he could not be neutral; either he destroyed, or he promoted. "Nature" could not guide vegetation toward the climatic climax, or repair damage, because nature, as action of a natural world independent of man, simply did not exist—certainly not in the midcontinental grasslands. The scientist had to discover by experiment or historical inference what the path of succession would be isolated from man and then actively guide the path of succession. This was called *conservation*. It was a vast lesson to learn, and learning it put Weaver in another world from the innocence of the Clementsian philosophy of the previous generation. Weaver in 1938 had moved as far from Clements's *Plant Succession* of 1916 as the adult who has discovered complicity has moved from the world of the child. The "structure" of the world may not have changed physically, but the vision illuminating it was significantly different. The second edition of *Plant Ecology* stressed repeatedly the necessity for man to act, rather than simply to witness. Things would not take care of themselves. "To remedy" the destructive coactions of man on environment, "it then becomes necessary to reverse both sets of processes [coaction and reaction] and turn them into agents for stabilization and control."[23] Or, addressing the problem of disruption of wild animal life, "destructive coactions must be replaced by constructive ones, and natural or artificial propagation must be invoked to restore a dynamic balance within species as well as communities, just as in the plant cover itself."[24] Cultivation, overgrazing, fire, erosion: "The damage once done, it can only be repaired by man turning to the use of coaction and reaction for constructive rather than destructive effects."[25]

The practical loss of vegetative progressivism, abandonment of the concept of the formation as an actual, individual organism, and the admission that man must play a role as conservator of the "natural" world were significant adulterations of the pure milk of Clementsian-ism. If Weaver and Clements, in their textbook, were so close to the positions of Clements's critics, we are justified in asking why this was not simply an acceptance of their critics' arguments. Is this not a good example of a scientific doctrine being rationally modified by experi-ments and arguments of opponents? I believe that it is not. Rather, the modification of their views represented an experience internal to their microparadigm, the struggle against the drought on the Great Plains. This experience was contrary to their assumptions, and this was a part of the reason for the shift in their views, as it was also a shift in their institutional roles.

It is not clear how much contrary experiments or arguments were capable of rationally convincing Clements that he ought to modify his views. We have already seen that he responded privately to Gleason's attacks with the dismissal that Gleason had not expe-rienced the prairies in the same way Clements had. In 1935, after Tansley's attack in the "Use and Abuse" article, Clements again responded by terminological shifts and by willfully believing that there was less of a gap between his and Tansley's views than Tansley said there was. His carefully considered response to Tansley was put through three drafts, the final one bravely asserting that "I do think that the differences are fewer and less serious than your style indicates, since some are more a matter of word preference than of actuality."[26] Earlier, he had written to Weaver that he did not think that man had had any significant impact on the climatic climaxes, a view strangely contradictory to his and Weaver's earlier work on *Experimental Vegetation*, in which they hypothesized that the dif-ferentiation of the postglacial grassland into short grass and tall grass prairies was a recent result of grazing. It would be also a contradic-tion of the views expressed under his name, as coauthor, in *Plant Ecology*'s second edition, as we have already seen, and expressed in 1936 in his *Environment and Life in the Great Plains.*[27]

The fact that Clements privately did not appear to modify his earliest published views leads me to suspect that the revisions of the second edition of *Plant Ecology* were made by Weaver, not Clements. Furthermore, I did not find in Clements's correspondence with Weaver any mention of revising, which there probably would have been had Clements contributed extensively to it. Clements was, at this time, occupied with the composition of the textbook *Bio-Ecology*, which he was coauthoring with Victor Shelford and which appeared in 1939, just a year after the second edition of *Plant Ecology*. Without manuscripts by Weaver, or oral testimony, we cannot be certain of this assumption, of course, but there is no evidence that it is wrong.

Since the retreat from pure Clementsianism in *Plant Ecology*, second edition, appears to have been the work of Weaver, we can look to Weaver's publications of the 1930s to determine the basis for the retreat. As we shall see, without any reference to the long tradition of criticism of Clements's views, Weaver straightforwardly responded to the Great Drought. Joining in collaborative research with Fred Albertson of Kansas State College, Fort Hays, Weaver published a series of reports on the drought. In 1936, they reviewed the two worst years, 1934 and 1935. Returning to the prairies they had previously studied in healthy condition impressed the scientists with the severity of the plant loss. Outside Lincoln, hilltop prairie had lost 45 percent of its basal plant cover in early summer 1935, with shallow-rooting plants of course suffering the greatest losses.[28] Patches of open soil appeared, and weeds penetrated into the once-impenetrable prairie sod.[29] At one prairie, the dominant prairie grass had been completely replaced by wheat grass, and in another, forbs had pushed out the little bluestems.[30] At a prairie southwest of Lincoln, Nebraska, the original climax prairie had been "all but destroyed."[31] Further to the west, where the drought had been more severe in preceding years, even ungrazed little bluestem mixed-prairie lost an average 88 percent of its bluestems.[32] The 1936 publication was an objective report by examining physicians, without speculation on the prospects for recovery. Perhaps they expected the drought to abate, but of course it did not for years and they would later be forced to wonder whether the native prairies could make a comeback.

After seven years of drought, significant structural changes had occurred in the grasslands. First, examination of several thousand square miles proved that mixed-prairie with distinct layers of mid-height and short grasses had been destroyed and replaced by short grass plains. Second, the surface soil had been so bared that weeds like the cactus had taken over the land and made agriculture impossible. Weaver and Albertson estimated that 20 percent of the soil was covered by the cactus (*Opuntia*).[33] Deterioration of the grasslands was so advanced that even occasional showers did not have their traditional rallying effect. Seedlings would sprout, only to be quashed by the heat.[34] The recovery mechanism of normal grassland cycles did not operate in the extremely pathological condition. Weaver and Albertson knew well the "almost incredible" recovery capacity of the ranges, but in 1940, they were forced to hint that the point had nearly been reached at which the original grasslands could not regenerate.[35]

The rains finally returned to the grasslands in 1942, but not in time to save the prairie. Weaver had to report that the true prairie of eastern Nebraska and Kansas had been destroyed and replaced by mixed prairie, just as the mixed prairie further to the west had earlier been supplanted by the short grass plains. This was more than the normal cyclical adjustment of the geographical boundaries of the types of grass communities. The previous dominants, little bluestems and big bluestems, had been eradicated completely. Wheat grass, which had a tough ability to exclude invading weeds, replaced the bluestems.[36] The grassland formation that Clements and Weaver had once described as the terminal climatic climax, in perfect harmony with the environment, was destroyed and replaced by a different set of dominants.

How could this have happened? Did man have a hand in it? Weaver stated in one article that the destruction of the true prairie and its replacement was a simple effect of the drought and would have occurred without weakening of the prairie cover by agriculture.

> Destruction of a portion of one plant association and its replacement by another has just been completed. This has occurred not as the result of man's interference with the vegetation nor the effect of his grazing animals. It has been due to a dry climatic cycle and has been accomplished within a period of seven years.[37]

Where agriculture had weakened the prairie, wheat grass had made an early entrance, but the drought persisted long enough for wheat grass to supplant native dominants even where agriculture was not developed.[38] That he was not entirely certain, however, that agriculture was unrelated to the devastation from the drought was clear in his famous article on the prairie in the *American Scholar* the following year (1944). After painting the beauty of the tall grass prairie, Weaver ended "The North American Prairie" with the dreadful lesson of the previous decade: man had the power to destroy the prairies beyond recovery and short of that only man's active management was hope for saving them. "[The prairie] is a slowly evolved, highly complex entity, centuries old. It approaches the eternal. Once destroyed it can never be replaced by man."[39] A decade earlier, in 1934, Weaver had raised the possibility, by quoting another scientist, that the true prairie could be preserved only by artificial control. The drought proved that, unless preservation accompanied agriculture, the true prairie and the mixed prairie might not survive. The expectation that the tall grass prairie would extend its range over the short grass plains, with which Weaver had begun his career, was a lost hope.

Knowledge and Ideology

The erosion of Weaver's faith in the postulates of Clementsian ecology involved three ingredients: the Great Drought's punishing denial of key Clementsian assumptions about the grasslands, recognition of the necessity of conservation practices to maintain climatically oriented succession toward the true prairie, and the grassland school's apparent insulation from long-standing criticisms of Clementsianism. These three ingredients effectively deny that this shift in assumptions and knowledge occurred in the manner prescribed by the traditional interpretation of the history of science. Weaver and the grassland school's understanding was not changed by rational argument of critics or by experimental disproof. Testing of Clements's views was certainly going on in the 1930s, but these tests did not show up in the grassland literature. In the entire run of citations of the bibliography of the school, there were few or no references to the critics. And the critics played no discernible part in Weaver's shift.

Recognizing the three ingredients of the shift does not tell us what the relationships among them were, or of their relative influences in initiating the shift. Without Weaver's manuscripts, unfortunately, these aspects of the shift probably cannot be known conclusively. We are left with the task of simply providing a reasonable explanation. I propose that in the 1930s Clementsian philosophy, as it was exhibited in the grassland specialty, behaved ideologically. According to Karl Mannheim's classic exposition of the sociology of knowledge, a person's "type of participation" in society determines the formulation of problems, the level of conceptual abstraction, the selection of factual content, and the recognition of the grounds of validity.[40] "There are spheres of thought," Mannheim believed, "in which it is impossible to conceive of absolute truth existing independently of the values and position of the subject and unrelated to the social context."[41] From this point of view, Clementsian philosophy is interpreted in the 1930s as responding to the mode of the grassland school's participation in the Great Drought. Although their professional positions since Bessey's era had included much social service, in the 1930s Weaver and his colleagues moved dramatically to increase the extent and change the character of their service as scientists in combatting the drought. Parallel with this increased social involvement, they discovered that the grasslands were not, after all, largely uninfluenced by man's interaction and that they could—indeed, needed—to intervene to manage the path of succession. In terms of chronology, the emergency of the drought brought them into conservation practice early in the 1930s and their attitudes fundamentally shifted, according to the evidence, later. Consequently, we would say that the participation of the scientists in fighting the drought led them to reinterpret nature in a way that made theoretically possible the role they had already assumed. Making their role theoretically possible required extensive modification of Clementsian philosophy. No doubt, the scientists did not self-consciously alter their knowledge to meet their social needs; had they done so, their scientific knowledge would have been only rationalization of social practice. Mannheim did not mean that knowledge was ideological in so crude a way. Rather, the structure of their knowledge rationally changed in response to the change in their social interest, and the rationality derived, not from the character of the knowledge, but from their interest. Rationality, in Mannheim's view, is a function of the person's social position and stake in society, that is, of the historical-social perspective of the person. So the changes in Clementsianism were rational to the grassland school because their needs and perspective had changed.

This interpretation of the shift in the Clementsian philosophy in the 1930s as ideological brings us back to the interpretation, with which I opened this study, of the origins of the Clementsian philosophy in the 1890s. I examined the origins in terms of the antipositivistic philosophies of science of Michael Polanyi and Thomas Kuhn. The Clementsian philosophy of ecology grew in a small community of shared experiences, training, values, and expectations (the "Bessey system"). Part of these experiences was of the grasslands themselves, an experience expressed theoretically only when contrary views of the German botanical geographer, Drude, forced their articulation. The approach to grasslands ecology involved a rigorous search for causality in the plant-habitat relationship, so that the scientist would have specific knowledge to bring to the problems of the farmers, largely concerning the level of productivity and protection of crops from pests. The origins of the microparadigm had been ideological in the sense of reflecting the needs and perspective of the Bessey group. Two generations later, the situation of the scientists had changed and the grasslands themselves were changing. The scientists were involved with wholesale salvage of the grasslands through application of conservation practice over vast areas. Earlier they had been a group of scientists offering a microparadigm in competition with an alternative microparadigm from Chicago; in the 1930s, they were the establishment, the competitors having withdrawn.

The involvement of the grassland scientists in conservation during the drought began at once. Clements saw the crisis as the chance to apply his ecological system on a wide scale. As he wrote to the Soil Erosion Service, "dynamic ecology has an exceptional opportunity here in connection with great public works for erosion and flood control."[42] In 1934 and 1935, Clements got his "boys" involved in several large federal projects, the controversial shelter belt program, the exposure program, and the revegetation of native grasses. He had been requested in 1933 by Dr. Charles Merriam, head of the Carnegie Institution, his employer, to place himself at the service of the agencies dealing with erosion problems on the plains. He was subsequently an official consultant to the Soil Erosion Service and, when this agency was combined with others into the Soil Conservation

Service in early 1935, he was automatically a consultant to the Soil Conservation Service.[43]

The shelter belt program originated in the late spring 1934, as a response by President Roosevelt to the political pressures on him to do something about the drought-aggravated depression in the Midwest. Drawing upon an idea presented in the election campaign of 1932, Roosevelt proposed a belt of trees to be planted from North Dakota to Texas. There would be one hundred "parallel strips, a mile apart and each seven rods wide."[44] Overall, 1,820,000 acres would be planted in eight to ten years. Presumably, the trees would reduce dust storming and increase moisture by decreasing evaporation. Farmers whose land was leased would receive a small cash payment. The project was, therefore, to provide long-term benefits to the plains environment and immediate economic relief to the distressed farmers.[45] Clements and other scientists and advisers, including Rexford Tugwell, doubted that any environmental benefits would result from the belt. But Clements was fascinated and supported it because it promised to be a marvelous scientific experiment.[46] Exactly what role Clements took in the following summer and fall regarding the shelter belt proposal is unclear. During the summer, federal forestry officials traveled in the Midwest, consulting with scientists and organizing regional federal offices. Clements corresponded with various federal officials and apparently met with touring scientists at his Alpine Laboratory in Manitou Springs, Colorado.[47]

At all events, the shelter belt program died in 1935 and Clements's interest turned to another project, the "exposure program." Clements wanted to test the carrying capacity of grazing ranges and to devise methods whereby the federal government could annually predict the carrying capacity of the public domain. Such an annual program would enable the government to prevent overgrazing and thereby reduce one of the major causes of the destruction of the grasslands. On the advice of Walter Lowdermilk, assistant chief of the Soil Conservation Service, Clements obtained support for the project from Secretary of the Interior Harold Ickes. Tests were begun in 1935 in California.[48]

Clements proposed to fence off large patches of grassland around the country, as test plots, thereby protecting them from rodents and grazing animals. These plots, ranging up to several hundred acres, were to be tested under a variety of grazing conditions. "Processes" such as "trampling, burrowing, erosion, burning, denuding, grazing, competition, seeding, and planting" were to be studied for their role in establishing the grazing capacity of the grass cover. The source of Clements's inspiration for the enclosure project was obvious: the meter-plot. Clements's test enclosures of hundred-acre size were merely enlarged quadrats. Indeed, Clements indicated in his proposal that the type of research to be conducted in the enclosures had been conducted for decades in transects and quadrats.[49] The project would take several years, to include annual precipitation and drought cycles. With this information Clements hoped that the government could classify the lands in the public domain according to their capacities and administer them so that the grass cover was protected while maximum grazing occurred.[50]

The University of Nebraska botany department's involvement with the drought and the general agricultural crisis of the 1930s included John Weaver's personal efforts, but extended also to other faculty. The department received a thousand dollars a month for at least two years (1934–36) to support fifty to seventy botanical students doing work related to the drought crisis. Faculty each received forty dollars a month for supervising the federally supported students. Understandably, the federally sponsored work and the necessity for finding research for students steered faculty toward drought and other practical problems.[51]

The preoccupation of the grassland school of ecology as a whole with the crisis is indirectly indicated by the career patterns of the fifty-eight scientists. The production of grassland ecologists increased at the University of Nebraska in the 1930s. From the coauthorship patterns of John Weaver with his graduate students, it is clear that much of the graduate work concerned practical problems relating to grazing, cropping, conservations, and erosion control. Toward the end of the decade, there was an increase in federal employment by Nebraska doctorates, and all during the decade there was a steady climb in university employment by the Nebraskans, undoubtedly reflecting the increasing importance of experiment station research.

The work of the grassland group in the drought crisis was at once both the victory of the Clementsian paradigm and its undoing. As Clements triumphantly and justifiably trumpeted in "Ecology in

Public Service" in 1935, mitigation of the effects of the drought depended upon Clementsian ecological theory of succession and climatic climax. But at the same time, this victory blunted incentives for the grassland group to try to direct Clementsian theory toward the newer theories in ecology. If it had been Clements's adamancy in his own rightness in the 1920s that prevented him from urging his disciples to head in new directions, in the 1930s, the new directions were also practically inconvenient. The grassland scientists had a vastly enhanced public role in conservation. They could not be both innovators in theory and conservationists in practice and they chose, as the bibliography of the group in the 1930s clearly shows, to be the latter.

The new role of the grassland ecologists included social planning. Clements envisioned, in its most naked form, that a plan to save the plains would require the federally regulated use of the midcontinental grasslands on scientific principles and the resettlement of the region's rural population. The mid-1930s were years of visionary social schemes, and Clements rose to the occasion. The ecologist was to be a social planner and manager, deriving his authority from the highest level of government. Tansley had not been wrong to sense that organismic ecology was tied to political principles, or that these principles led to the authority of the largest unit of society over the individual. Clements laid out the matter clearly to Lowdermilk, the associate chief of the Soil Conservation Service in 1935:

> It is most important to consider the restoration of the grass cover in the light of the economic and social values to be sought for the region as a whole. The regeneration of the grazing ranges is a relatively simple matter in comparison with these, but to be successful in human terms it must be considered in relation to desettlement, land classification, and resettlement. Moreover, the latter must be something far more intelligent than the original trial-and-error settling-up of the country.[52]

It is little wonder that Clements in 1935 publicly quoted with approval Jan Smuts as saying, " 'Ecology must have its way: ecological methods and outlook must find a place in human governments much as in the study of man, other animals, and plants. Ecology is for mankind.' "[53]

Clements believed that he was on the verge, in the mid-1930s, of tying the climatic cycles of rainfall and drought to sunspot cycles. If this effort, which had had a somewhat questionable scientific legitimacy for several decades as handled by researchers like Ellsworth Huntington, were successful, then the new power of prediction would require a thorough overhauling of America's social and political systems.[54] The government would have the ability to predict, a decade ahead, the impending weather catastrophes and would need the power to prepare society to meet them. Clements thought that the political consequences of such predictive power were so obvious that they did not need mentioning. "It is unnecessary to point out the significance of long-range prediction for the various projects previously discussed [e.g., land reclassification], as well as for the whole social-economic system."[55]

The chief theoretician of grassland ecology moved easily into the role and ideology of social planning. His project for ecological land classification implicitly criticized a century of national land policy, and he claimed his criticisms of the "trial-and-error" and individualistic settlement of the grasslands went back to his early publications.[56] As the previously quoted letter to Lowdermilk indicates, he recognized that his ecological rescue of the grasslands required profound social and economic reconstruction. He provided an ideological statement of this reconstruction in the closing pages of his work, coauthored with Carnegie Institution paleontologist Ralph Chaney, *Environment and Life in the Great Plains* (1936). He proposed that the individualism of American midland agriculture be ended and farms and environment be synthesized into a large, organic whole, coordinated by the federal government. To understand the ideological union between scientific ecology and social reform, we must look at several long quotations from Clements's work.

> The need for ecological synthesis is but half met by its application to the processes of recovery. It must also be accorded the fullest expression in the organization of farm and ranch, as well as other units in practice. Each of these must be reshaped into an organic entity, with all the parts present and coordinated to bring about optimum economic and social results.[57]

Clements perceived traditional American individualism as an obstacle to his program. Federal agencies, such as the Soil Conservation Service, were but first steps in the direction of national coordination.

His statement of support for a New Deal in ecology may have been vague in detail, but its ideological commitment was clear:

Coordination of process and practice on farm or ranch must be reflected in the organization of the community. This signifies cooperation, a community function still almost undeveloped except with respect to marketing. Though by far the most important of social processes, it has made such slight progress against myopic individualism as to confirm the belief that, like the functions of other and simpler organisms, it can be evolved only under the stimulus of outside forces. Fortunately, times of stress provide the very pressure needed, as well as the agencies to guide the response to it, and it now seems probable that cooperation will be set forward more in the present generation than in the full century since the settlement of the West began. However desirable it may seem that this and other social functions make a natural growth, it is evident that the nationwide experiment of having the first steps directed by the Soil Conservation Service, Forest Service, and Division of Grazing in their various jurisdictions promises the only adequate solution of this crucial problem in social progress. To the ecologist, who recognizes that society as a complex organism is certain to evolve in harmony with its environment, it is of critical importance that the environment be fashioned as to call forth progress and not retrogression.[58]

The political implications of the manifesto quoted above could be inferred without difficulty. Democratic political processes, based on voter participation, were obstacles to progress, because they allowed conservative individualism to assert itself. Administrative organization of society by federal agencies, staffed by scientifically trained managers, was necessary to solve the social problem of economic depression and misuse of the land. Did this mean that the individual and the small social unit were less important than the whole of society? Or that local authority over life had to be submerged under national authority? Clements did not shrink from these conclusions:

What has been said of farm and community applies with like force to every social integration of higher rank, and especially to the nation as a whole at a time when it has delegated powers to a degree hitherto unknown. Within such a huge organism, the whole is much greater than the mere sum of its parts, and hence the need for coordination and correlation far transcends all other considerations whatsoever. It is here that cooperation will meet its supreme test, and it can emerge from this successfully only as the ecological ideal of "wholeness," of organs working in unison within a great organism, prevails over partial and partisan viewpoints.[59]

The Madisonian vision of American society, which had guided the framers of the Constitution and spoke through the Federalist papers, disappeared under the siege of Clementsian organicism. A society based on individual competition and political adjustment of conflicting interests had despoiled the grassland and turned a drought into a disaster. Politics must give way to scientific management of society.

The themes of Clements's ideology were not new. The idea that the complexity of modern society, including mechanized farming, made nineteenth-century political institutions antiquated was central to the progressivism earlier in the century. Similarly, the call for scientifically based management to replace political processes had been made by professional classes for a generation. Many federal scientists, from John Wesley Powell to Clements's friends in the federal erosion, forestry, and conservation agencies, had argued that the special conditions of the semiarid West made necessary federal regulation of land use. The call for a more scientifically rational use of the public lands and a reorganization of midwestern society to fit that use were made in the 1930s by other persons besides Clements, including his friend, Walter Lowdermilk, assistant chief of Soil Erosion Service and Soil Conservation Service, Hugh Bennett, chief of the services, and Henry A. Wallace, secretary of agriculture. Ideological expansion of old ideas was a feature, albeit often controversial, of the Roosevelt New Deal. At the same time that Clements was thinking through the scientific exploitation of Roosevelt's shelter belt of trees, the federal commissioner of reclamation, Elwood Mead, was calling for the voluntary abandonment of marginal farmlands, aided by federal incentives. Harry Hopkins's Federal Emergency Relief Administration set up "Rural Rehabilitation Divisions," whose purpose was to resettle farmers on relief onto new farms. Walter Lowdermilk later toured Palestine for the Department of the Interior, to study agricultural and conservation practices in a region climatically similar to the U.S. semiarid West, and returned favorably impressed by the cooperative settlements and national authority for directing development.[60]

The official position of the Roosevelt administration on the social and economic issues raised by the drought was presented in the report of the presidentially appointed Great Plains Committee. This committee had been appointed in September, 1936, with Morris L. Cooke, the prominent liberal and director of the Rural Electrification Administration, as chairman. In the committee's report, "The Future of the Great Plains" (February 1937), the views of Clements, and undoubtedly many other concerned experts, were legitimated, though their radical quality was not denied. The report stated straightforwardly that the destruction of the native and cultivated vegetation of the Great Plains was the "result of human modification of natural conditions."[61] The ecological balance of the plains that existed before agricultural settlement had been destroyed by inappropriate cultivation practices. American settlers from the Mississippi Valley assumed that they could transfer high-rainfall agriculture directly to the plains, because they settled the plains in the post-Civil War era at a time of higher than normal rainfall. Return of the normal rain cycles led to agricultural failure and destruction of the land at the end of the nineteenth century. The drought and ruin of the 1930s was but a continuation at crisis intensity of a basic situation that had existed for two generations.[62] The only permanent solution to the problem, which included economically and ecologically viable agriculture, was depopulation, resettlement, restructuring of the farm size, and new cultivation practices. Despite the best efforts of the Great Plains Committee to make the solution appear as no break with traditional American values, the solution in effect required federal guidance of the development of a region and suspension of individual decisions and many state powers.[63] Clements's correspondence and *Environment and Life in the Great Plains* made clear that he endorsed many of the recommended mitigation measures as the only steps that would scientifically meet the crisis.[64]

The writings of the secretary of agriculture, Henry A. Wallace, provided an ideological framework to fit around the land program of Clements and the plans of agencies in his department. It was in Wallace's department that Congress in 1935 located the newly created Soil Conservation Service. Wallace's *New Frontiers* (1934) appealed for broad support for the New Deal in agriculture and land policy. The machinery of free competition of nineteenth-century American society, by which economic goods were supposedly equitably distributed, had been antiquated by the rise of the large industrial corporation and the concentration of economic power in the hands of a small number of corporations. The ethic of individualism and small government, which was appropriate to the bygone era, inhibited solution of the crisis of the depression. Wallace, like most New Dealers, believed that governmental administration of certain sectors of the marketplace was necessary to raise the economic condition of disadvantaged groups, such as farmers. Prices in these sectors had to be set by group planning, rather than by competitive bidding. The secretary of agriculture believed that only loose federal planning was necessary for most sectors of the economy, with the important exception of natural resources where detailed plans were needed. Consequently, use of the national domain was to be carefully regulated. He applauded the creation of forestry reserves, beginning with the Theodore Roosevelt administration. Although the Taylor Grazing Act had not been passed when he wrote his book, Wallace was in favor of its regulation of grazing on the public grasslands.[65] "In many parts of our social structure the blueprint method of approach is not advisable, but land is so fundamental and precious a heritage, that we should outline a policy to continue over many administrations, and stick to it for the sake of our children and their great-grandchildren."[66]

Wallace recognized that the New Deal offered a new ideology, public planning, and *New Frontiers* was a straightforward political appeal on behalf of that ideology. We must recognize that Clements's call for scientific administration of the natural environment, on ecological principles, was similarly ideological. It was intimately related to the political changes of 1934 and 1935. Not simply did New Deal programs and politics lead Clements to formulate his view, not simply did the extremity of the midcontinental drought demonstrate the inadequacy of historical federal policies in the grasslands, but the ideology of the New Deal provided a new means for academic scientists to conceive of their role in American society. By 1935, Clements perceived the ecologist as a planner whose responsibility must include the private sector, as well as the public. By 1943, Weaver had come to perceive also that maintenance of the native grasslands required active management by the scientist as conservationist.

Intellectual reformulation of the Clementsian microparadigm accompanied the new conceptualization of the role of the grassland scientist. Before the 1930s, Clementsian ecology was based on assumptions that vegetational change was inevitably progressive and that man's role was only destructive. By the mid-1930s, Clements had come to feel that man's active role in the management of nature was necessary. While verbally he still defended the notion that nature independent of man was progressive, in practice he recognized that nature did not so exist. Weaver did not make the obvious ideological pronouncements that Clements did; nevertheless, his publications in the 1930s and early 1940s revealed a clear backing away from pure Clementsianism. He abandoned the concept of an organism, of inevitable progress, and of inevitable recovery. These intellectual changes came after the scientists had begun their participation in the struggle against the drought in the Great Plains. We naturally see the intellectual changes in Clementsian philosophy as resulting from the new understanding of the interaction of man and nature provided by the scientists' new management role. For the changes they made in their ecological theory, as we have seen, not only allowed a management role for the scientists, but made that role necessary to preserve the native grasslands. In this way, the changes in Clementsian ecology in the 1930s reflected the changed social role and status of the scientists sharing the microparadigm; this is what I mean when I say, in the spirit of Mannheim, that scientific knowledge was behaving ideologically.

Exhaustion of the Research Specialty

World War II brought significant changes to the scientific specialty of grassland ecology. The collective biographies show a rapid decline in degrees granted during the war; indeed, only two doctorates were awarded in the years from 1941 to 1946 to subject scientists who would later contribute to the grassland bibliography. This compares to six doctorates awarded in the preceding six years (1935–40) and seven doctorates in the following six years (1947–52).[67] The rate of publication declined slightly. The mean number of annual publications in the years from 1935 to 1940 was 17.5 ($\sigma = 6.4$); the mean from 1941 to 1946 was 16 ($\sigma = 4$); from 1947 to 1952 the mean was 15 ($\sigma = 4.4$). The significance of the decline was, of course, simply that it was a decline. Prior to 1941, the rate of publication had been steadily increasing.[68] Finally, the number of grassland scientists employed in scientific work declined after 1941. The number of federal grassland scientists declined from seven in 1941 to four in 1945. The number of university and experiment station scientists declined from twenty-two in 1941 to nineteen in 1945. The decline in university employment had begun in 1939, however, from a high of twenty-eight scientists in 1938.[69]

The drought only coincidently ended the same year that the United States entered the war, but the coincidence reinforced the decline. The emergency of the drought stimulated employment of grassland scientists and justified their social role. Their research swelled in response. For Weaver at Nebraska, Albertson at Fort Hays, Kansas, Hanson, who had just left South Dakota, and the other scientists, the shift from 1941 to 1942 was abrupt. Their research mission and the nation had suddenly entered a new era. In 1942, grassland ecology advanced to old age. Both the theoretical development and the practical testing of the specialty's microparadigm lay behind them. The great national context of the New Deal and emergency that focused attention on them was gone. The leaders of the struggle against the drought were, in 1932, at the peaks of their professional careers, with a mean age of 47.9 years. Ten years later, they looked toward the end of their careers. In 1932, Clements was 58, Weaver 48, Albertson 40, Hanson 42, Pool 50, Shantz 56, Sears 41. With the exception of Sears, who was making a distinguished career in forest ecology, the struggle with the drought was their last great research effort in the grasslands. Hanson, for instance, published six papers in basic research on the grasslands during the drought period, but no similar publications between 1939 and 1950. Albertson published seven basic papers between 1936 and 1942 (seven years) and five between 1943 and 1953 (eleven years); this represented a decline by one-half in his rate of publication. Sears published *Deserts on the March* (1935), then turned away from grassland studies until the end of his career, when he published *Lands Beyond the Forest* (1969).[70]

The decline of grassland ecology as a specialty can be interpreted from different perspectives. First, sociological aging is apparent in the changes discussed above. It is inconceivable that the specialty would not slow down or dramatically change in character as its first and second generational leaders passed through their own middle age, their great accomplishments behind them. But this is hardly a sufficient explanation. Graduate training continued; why should not a new generation of leaders have arisen to invigorate the school? The decline must have also involved a theoretical enervation to produce the decline of publication in basic research. Winning the drought and assumption of their role as conservators cost heavily the theoretical power of the Clementsian microparadigm. By 1938, as we have seen, the grassland scientists had drastically restricted the extent of generalization of Clements's major principles. Did they hope in the 1940s to recoup what they had earlier lost in theory? By then, experimental disproof of Clementsianism was beginning to come in from the critics. Further, the rapid development of the synthetic theory of evolution, which involved an ecological framework for genetics, and the increasing understanding of the biochemical basis of inheritance made Clements's theory seem antiquated. Theoretically diminished, Clementsian ecology did not have the intellectual potential of countering the most dazzling triumphs of twentieth-century biology.

The different sociological and philosophical interpretations of science, which I have utilized in this study, provide alternative scenarios for the decline of a scientific specialty. But, surprisingly, these scenarios must be inferred from the sociological theories, because sociologists have primarily studied the rise and establishment of scientific groups, not their decline. Apparently, historians have the idiosyncracy of also asking questions about the decline and fall of the powerful. Thomas Kuhn, in his theory of noncumulative scientific revolutions, did not see specialties as "declining," in the sense of gradually decreasing activity in research, publication, and graduate training. Rather, specialties were abruptly abandoned during a theoretical crisis, with scientists not converted to a new view eventually, and literally, dying off. The new group of scientists took up a new microparadigm as an alternative to the established microparadigm, basing explanation around the phenomenon that had been anomalistic in the old microparadigm. These new scientists lacked the special training and experiences that enabled holders of the old microparadigm to be *un*troubled by its anomalies. The sociological result of the theoretical shift was the rapid replacement of one group of practitioners by another.[71]

Another scenario for the declining era of a specialty has been provided by Diana Crane. She suggested simply that the old microparadigm was theoretically exhausted, without necessary reference to a new competitor.[72] What Crane meant by "theoretical exhaustion" was by no means clear, however. The term implied that a theory had a limited power that could be expended. It is as if a theory could only generate, say, a hundred predictions, and during the time that scientists were making the first fifty predictions the specialty would rise, and in making the last fifty, the specialty would decline. Presumably, as scientists saw the approaching end, they would try to shift to another theory. This concept implies that scientists can behave toward the impending end of their microparadigm as if they already possessed hindsight on its termination; of course, that is not possible. While scientists frequently have a feeling of the predictive capacity of a theory, if that capacity is limited, they will generally try to ascribe that limitation to their own development of the theory at the moment, not to the theory objectively. As Kuhn has said, in analogy to a cliché, it's a poor scientist who blames his tools for his poor performance. The notion of theoretical exhaustion does not, therefore, help us to explain the decline of a specialty from the point of view of its practitioners.

Recently, Wolfgang Stegmüller, a philosopher of science at the University of Munich, has provided a theory of the decline of a specialty that analyzes the kinds of intellectual changes a theory undergoes in decline. His interpretation is basically Kuhnian, but it provides an explanation of gradual decline of the sort the grassland school underwent after 1941. Stegmüller interprets the abandonment of a paradigm as a process of intellectual restriction of empirical claims, rather than simply rejection *in toto*. As this restriction takes place, the community of scientists is divided into several groups: the original practitioners who shared in the creation of the original "paradigm example," and other scientists who were brought to the microparadigm when the original scientists extended their empirical claims to new paradigm examples. For this latter group, rejection of the microparadigm tends to be abrupt, leaving the original practitioners behind, gradually whittling down their microparadigm.[73]

Stegmüller's theory depends upon his characterization of the intellectual contents of a microparadigm. While I cannot judge the adequacy of this characterization for all specialties, the Clementsian microparadigm clearly meets his description. He distinguishes between two components of the microparadigm, the theoretical core and the "intended paradigm examples," or applications. The theoretical core consists of three parts: an axiomatically linked set of defined entities, which constitutes the subject matter of the paradigm; various models of the applicability of the theory to the world; and functions that link applications. In Clementsian theory, the basic defined "entity" of the theory was the formational organism; this was the natural object that the theory concerned and which scientists studied. It was axiomatically linked by Clements to other entities, including the "seres," or organic stages in development of the formation. Historically, the original "intended paradigm example" was the North American midcontinental grassland, and Clements quickly picked up alpine forests as a second intended paradigm example. Whenever Clements or Weaver wished to provide basic illustrations of the application of Clements's theory to nature, they turned to the grasslands. Other applications were made in analogy to the basic example. Whenever grassland ecologists felt their microparadigm challenged, as by the European plant sociologists in the late 1920s and 1930s, they instinctively turned back to the grassland model for reassurance of the validity of their theory. Finally, with regards to Stegmüller's "functions," as a component of the theoretical core, Clements stated that succession was always and everywhere progressive. This verbal proposition acted in analogy to mathematical functions to which Stegmüller refers in the relationship between the defined entities of ecology. The plant formation always developed toward an adult form (the climatic climax). This was therefore a universal function of vegetational change. The function was derived inferentially from the concept of the formation as an organism. The formation could no more grow backward than a tree could reverse its growth and become a seed-leaf.

In Clementsian ecology, the intended paradigm examples were, as we have seen, the English landscape, grasslands outside of North America, and alpine forests. These paradigm examples were developed by a group of scientists, of whom A. G. Tansley, of course, was the most well-known European. The intended examples were sets of empirical propositions about the course of vegetational change. When scientists had difficulty in understanding a particular paradigm example, they referred to the original example of the North American grasslands.

Theoretical exhaustion occurs, in Stegmüller's scenario, by the invalidation of the intended paradigm examples. As these applications are disproved or rejected, the empirical scope of the functional operators is decreased, and the model of applicability is restricted. If this process should continue, eventually the microparadigm would consist only of the original paradigm example and a theoretical core that related only to it. For the original practitioners of the microparadigm, believing in the original paradigm example alone would adequately constitute "holding the theory," but for scientists brought later to the microparadigm by work with the other intended paradigm examples, this would not be satisfactory and they would "reject" the theory. Returning to discussion of Tansley's advocacy of Clementsianism, we saw that Tansley's acceptance of the theoretical core of Clementsian ecology was conditioned by his hesitancy about accepting the English landscape as an application of that theory. What came first, before the 1930s, was his hesitancy about the intended paradigm example, not his hesitancy about the theory; he was willing to accept the concept of the formational organism, as a heuristic model, for decades before finally rejecting it. Thus, as we have seen in this chapter, Weaver's acceptance of Clementsianism was weakened when experience with the drought contradicted the applicability of the theoretical core in its strongest form. The "strong version" of holding a theory involves, according to Stegmüller, holding on to the intended paradigm examples as well as the original paradigm example; the "weak version" of holding a theory means holding on only to the original paradigm example. Tansley and Weaver both, before the 1930s, held strong versions of Clementsianism. Tansley's rejection of its application to England, pressured by the political implications of ecology in the 1930s, led to total rejection of Clementsianism. Weaver, contradicted by the drought, fell back to a weak version. And Weaver's position in 1938 was indeed weak. He had come to deny inevitable progressivism and the reality of the formational organism. Clementsian ecology in the late 1930s had become "theoretically exhausted" in the denial of many intended paradigm examples, and in the restriction of the functional operators and denial of the reality of the primary defined entity in the theoretical core.

Deprived of so much predictive power and its forceful optimism, and centered by the drought experience on the managerial aspects of the causal relationships in succession in the grasslands, Clements's theory degenerated into technology. The weak version of the theory looked very much like the ad hoc understanding of the Great Plains that the Bessey group had achieved in the late 1890s, before Clements's generalization of it into a logical theory. The grassland school was moving, in the 1940s, from grassland ecology to range management.

Doctoral training reveals this shift clearly. In the decade preceding the drought, two of Weaver's three doctoral students derived thesis problems from the Clementsian microparadigm, but beginning in 1934, most thesis problems sprang directly from the effect of the drought on vegetation. Even as late as 1950, John Voigt's thesis was concerned with the composition of pasture in a prairie that had recovered from the drought that ended eight years before his research. Weaver directed or jointly supervised fifteen doctoral researches of grasslands scientists after the initiation of the drought in 1933. Nine of the theses directly concerned the drought or management problems of grasslands; another three indirectly concerned the drought or management, by providing information that could specifically be used to mitigate the drought or facilitate crisis management (as of drought-impacted and overgrazed pastures). After 1945, even archival citation of Clements's *Plant Succession* by these students ceases. One thesis that did refer to Clements's masterpiece did so to state that Clements had shown how succession was useful to range management.[74]

In summary, the research front of grassland ecology collapsed after 1941. The abandonment of fundamental Clementsian principles, the end of the drought, the sociological aging of the leadership, the decline in publication rate, the decline in graduate training, the interruption created by the Second World War: all suddenly and effectively coincided to break the pace of research established in the 1930s. The shift in citation patterns in grassland publications after 1941 dramatically illustrated the accumulated changes in the specialty. The "research front," symptomatically revealed by the citation of publications with a lag, or age, of two years or less, fell apart in 1942. In the years from 1932 to 1941, more than 25 percent of the citations in grassland publications were to titles less than three years old. In the years between 1942 and 1955, only 15 percent of citations were similarly to titles less than three years old. The statistics show the collapse in two ways (see Appendix table 18, Movement of the Research Front). The lag percentages within the bold lines from grouped years 1920 to 1924 to grouped years 1935 to 1939 show that higher than normal percentages of citations were made to recent titles. In the grouped years 1940–44, this pattern disintegrated.

The intellectual shift manifested itself professionally in the creation of a professional occupation, range management, with its own society and journal. A survey conducted in 1946 among "range men" demonstrated strong support for a professional society of their own, and a recruitment drive the following year signed up 500 members. So in January, 1948, the American Society of Range Management, devoted to "advancement in the science and art of grazing land management" and conservation, held its first meeting in Salt Lake City.[75]

The scientists of the grasslands generally and Nebraska researchers were well represented among the first generation officers of the new society. B. W. Allred, who took his B.A. and did graduate work, without taking the M.A., at Nebraska, L. A. Stoddard, who took his doctorate from Nebraska in 1934, and David Costello, a Chicago doctorate of 1934, were on the council (which had only six members). Fred Albertson was a member of the Organization and Policy Committee. Leon Hurtt (Nebraska B.A., 1914; no graduate degree) and Joseph Robertson (Nebraska Ph.D., 1939) were on the Nomination Committee. E. J. Dyksterhuis (Nebraska Ph.D., 1945) was chairman of the Elections Committee. David Savage (Montana State College B.A., 1924) was chairman of the Finance Committee. Of course, this strong representation ought not to be unexpected; the grasslands scientists were the most productive basic researchers in the field, and their names were sufficiently well known for nomination to society offices. Eventually, twenty-two of the subject grassland scientists joined the Society of Range Management, clear testimony of the shift of the profession toward applied or economic ecology. These scientists included the majority of the subject grasslands scientists active after World War II.[76]

The Great Drought of the 1930s had figured directly in the formation of the American Society of Range Management. The drought brought forth the Taylor Grazing Act of 1934, which harnessed grasslands ecology to range management. The Grazing Service in the Department of Agriculture was established. In 1936, the Agricultural Adjustment Administration initiated range conservation. The federal government's responses to the emergency of the drought had created a new profession out of governmental resources scattered throughout different agencies.[77] Understandably, scientists associated with the national government comprised most of the first year's membership in the new society.[78] Setting aside the war years, when professional organizing would have been forestalled, the Society of Range Management was established within seven years of the national government's massive intervention in the drought. The formation of the new range management society revealed the extent to which Weaver's shift had been part of a vast realignment of the intellectual resources of the national government and land-grant universities.

Epilogue

Full cycle: The school of grassland scientists grew, prospered, and declined, from the first planting on the virgin cultural soil of the Nebraska prairies to the final harvesting of technological benefits during the crisis of the drought. The father of the specialty's paradigm did not witness the denouement of his legacy. Frederic Clements died in Santa Barbara, California, in 1946. By then, his axiomatic system of plant ecology was in disarray, his secondary followers shifting away in the changing scientific theory of the postwar world. Only the primary advocates, established in their native prairie demesne, held onto the weak version of their theory.

Postwar opposition to Clementsian theory struck out from the individualistic approach of Clements's early critic, H. A. Gleason. Rather than conceiving of vegetation in terms of typologically classified structures, the new ecologists examined the individual plants as distributed on the landscape in a continuous gradient. Robert Whittaker, as a student working under Arthur Vestal at the University of Illinois, and John T. Curtis and his students at the University of Wisconsin, abandoned discrete vegetational zones and communities in studies of mountains and plains. These scientists were "new men." They began their work as students without prior professional commitment to the Clementsian paradigm. The passage from the old to the new intellectual modes was not conducted by the conversion of the primary advocates of Clementsianism, but by their succession by other scientists, who were outside the original matrix of experience that had given Clements and the early Nebraskans their convictions about the natural world. In the 1950s, heavy funding of ecological research by the National Science Foundation and the Atomic Energy Commission armed the new ecologists with sophisticated research technology. As one of the scientists instrumental in the new doctrines testified, "old-fashioned car-window ecology" was replaced by airplanes and helicopters; and radioisotopes and chromatographs replaced the ruler and string that had marked out the meter-plot. The pioneer days—we see the image of Pound and Clements riding their buckboard pulled by their mule "Moses" outside Lincoln, and of Edith and Fred digging "Billy Buick" out of the mud—were over.

METHODOLOGICAL APPENDIX

The Grasslands Bibliography

It is impossible to establish a comprehensive bibliography of publications on grasslands ecology in the test period (1895–1955). The central bibliographic source, *Botanical Abstracts* (1918–26, merging with *Biological Abstracts*, 1927 to present), only covers part of the period. No abstract series is inclusive. The *Botanical Abstracts* and *Biological Abstracts* especially failed to include the multitudinous publications of the United States Department of Agriculture and the semipopular monographs of the agricultural experiment stations, although these were regularly used by grassland ecologists. Moreover, since both *Abstracts* depended in part upon authors to send in notices of their publications, the appearance of an abstract often lagged several years behind the date of publication of the title. This lag lessened after World War II. The *Botanisches Centralblatt*, which began publication in 1880 and continued in the old series through 1919, similarly failed frequently to notice publications from the Department of Agriculture, the experiment stations, and occasionally some state academy of science publications.

My procedure in obtaining titles in the *Abstracts* was to pull out all titles, whether articles or books, listed in the *Abstracts* under the keywords "ecology" and "grassland," which concerned the midwestern United States and which were not economic in character. I also needed to find a source of titles for the early period before the abstract service. I was unable to find a comprehensive source and settled on the device of utilizing three nonsystematic bibliographies: James Malin, *The Grassland of North America* (Ann Arbor: Edwards Bros., 1947), John E. Weaver, *North American Prairie* (1954), and John E. Weaver, *The Grassland of North America* (1956). These three sources listed many titles not in the abstract series.

I checked the representativeness of the nonsystematic bibliographies for the period before 1918 with two techniques. First, I checked the list of titles in my bibliography against the list of titles in the Subject Index of the catalogue of the John Crerar Library, Chicago, which is considered a major repository of American botanical publications. My bibliography was so much more inclusive that the Crerar Index was of no value. I also checked the bibliography of titles for the years between 1880 and 1919 against the *Botanisches Centralblatt*; this important series for botanists had only 63 percent of the titles on my list. Second, I plotted the titles in my bibliography in a graph of cumulative publications and checked this against the logistic curve that Crane and Price believe is typical of the growth of scientific literature in this century. As I note below, my empirical curve is a close fit to the Crane-Price logistic curve for both the pre-1918 and post-1918 periods. I conclude that my nonsystematically collected bibliography is representative of the literature of its periods.

The resultant bibliography contains 535 different titles by 372 different authors. (A computer printout listing of the 535 titles is available by writing to the author.) These statistics of the bibliography are comparable to other scientific specialties (App. table 1).

When the cumulative bibliography of grasslands publications is graphed, the approximation to a logistic curve is immediately apparent (fig. 3 in the text). The curve moves chronologically through the expected stages: linear, exponential, linear, flattening. The chronological boundaries of these four stages can be determined in two ways. According to the theory of the curve, the boundaries come where the slope of the curve changes in growth rates, from linear to exponential and from exponential to linear. Using regression analysis, the chronological boundaries were thereby established: at 1916, as the last year when growth was closer to linear than to exponential, marking the end of Stage I; at 1941, as the last year when growth was more exponential than linear, marking the end of Stage II; at 1950, as the last year of linear growth. Nineteen-fifty-five is the last year of the bibliography; no publications were listed in *Biological Abstracts* for grassland ecology for 1956 and 1957. These periods were then tested by regression analysis to determine whether a linear or exponential growth rate best fit each period. The test conclusively confirmed the periodization (App. table 2).

Regression analysis of the curve of growth of the cumulative bibliography will not strike most historians as an appropriate means of deciding the periodization for historical development of the specialty. As historians, we would rather ask whether anything happened historically to provide chronological boundaries. We are ultimately after a theory of history, not a theory of bibliographies, and the latter is useful for us only if it aids in thinking through the former. In this case, however, the bibliographic analysis has pointed to years that appear to be important socially and intellectually in the development of the specialty. Nineteen-sixteen was the year of the publication of Clements's *Plant Succession*, which was one of two works in the grasslands bibliography published that year. According to Crane's theory, the microparadigm for the specialty should emerge out of the competition by the end of Stage I. Clements's big work of 1916 was recognized as the basic synthesis of the specialty's ideas. The analysis of the bibliography therefore supports the idea that *Plant Succession* was the paradigm-setting work for the specialty; following its publication, the bibliography of the profession expanded rapidly in the manner of normal science. We cannot simply attribute the increase in publication to the appearance of Clements's book, however. For the Ecological Society of America was founded in December, 1914, and began publication of *Ecology*, its professional journal, in 1920. From these two signal events, we would expect that earlier support of all specialties in ecology would be strengthened. There should have been an increase in publication in the years from 1914 to 1920, in other words, for historical reasons. If these events prevent us from authoritatively asserting that Clements's *Plant Succession* alone set off the increase in grassland ecological publication, they do reinforce the reasonableness of the discovery that publication rate did increase dramatically between 1914 and 1920, with these years marking the end of one era in the history of grassland ecology and the beginning of another.[1]

Events likewise provide independent confirmation of the idea that 1941 was the last year of a stage in the history of the specialty. At the end of the year, the United States entered World War II, which—we would expect—decreased the number of scientists and the funds for publication, thereby depressing the growth curve of the cumulative bibliography. Nineteen forty-one was also the end of the drought on the Great Plains, which had occasioned an extensive scientific literature. The return of the normal rain cycle removed a major phenomenon that the specialty had studied. The decline in the publication of empirical reports, though not necessarily of reviews and analyses, was reflected in the decline of the publication rate in the cumulative bibliography.

The periodization of grassland ecology is dramatically illustrated by the absolute frequency of publications (App. fig. 1). The number of publications in 1917 was triple that of 1916, and the number of publications of 1942 was one-third of 1941. (See also App. fig. 2, which provides a three-year moving average of the absolute growth rate of the cumulative bibliography.) Except for a brief increase in publications in 1950, the specialty was in general decline after 1941. Explanation for the publication increase of 1950 is not obvious, but examination of the institutional history of the specialty will show that following World War II, there was a brief increase in doctoral training in the field, presumably of returning veterans. In general, however, they did not remain in the field, but moved quickly into economic areas, such as range management.

Analysis of time series is always difficult, but the obvious importance of the chronological boundaries established by regression analysis of the growth curve should reassure us that the cumulative bibliographic curve does represent stages in the social growth of the profession. The theoretical curve, as used by Crane, can be seen simply as the growth curve of literature that would occur under

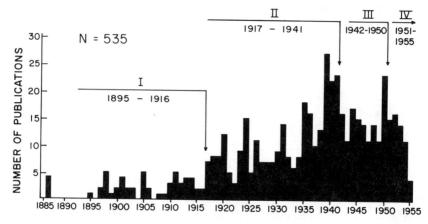

App. fig. 1. Absolute frequency of publications

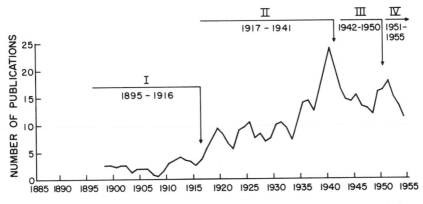

App. fig. 2. Three-year moving average of absolute growth rates of cumulative bibliography

optimal conditions of institutional growth. Scientists could behave solely out of considerations of knowledge and their own social interests; there was sufficient money that all graduate training, research, and publication needs would be met. Like economic supply-and-demand models, it cannot perfectly predict the empirical situation. But if we assume that scientists tended to behave as if the model were true, then deviations from the model would refer to large historical circumstances outside the value system of the scientists. Thus, the decline after 1941 was greater than the model predicts, but is easily explained by the initiation of World War II. Another deviation of the empirical curve of figure 3 from the theoretical model of figure 2 occurs in 1925, when the absolute rate of publications dropped at a time the model predicts it was rising (see App. fig. 1). While this pause in the growth rate may have no reason other than the rhythm of research and publication, it is also possible that it reflects a decline in federal employment of scientists. (See text figs. 5, 6, and 7 on doctorate employment patterns.) The precise cause of the decline of employment of federal scientists is not clear. Although the Department of Agriculture's budget was low during the early years of the Coolidge administration, the research budget of the department increased slightly during the middle 1920s. While we do not know how the funds were distributed, as between salaries and support, for instance, we should not hastily assume that grassland scientists were pushed out of federal employment for budget reasons. As with the case of the decline after World War II, the decline of the mid-1920s is

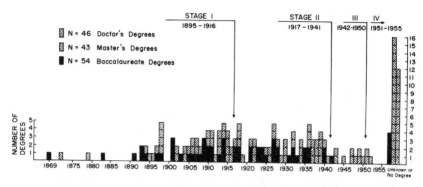

App. fig. 3. Absolute frequency of academic degrees, 1869–1955

explained by the collective biographies of the publishing scientists, which reflect the larger exigencies of the social world.[2]

The Collective Biography

Standardized biographies were compiled on all authors who published, as author or coauthor, three or more articles in the grasslands bibliography of 535 titles. Fifty-eight scientists met this criterion. The threshold requirement of three titles was based upon a historical judgment that practicing membership in a scientific community required more than one or several publications. I wanted a criterion that would eliminate those scientists whose only contribution was the doctoral thesis, carved into a couple of articles. For most scientists, collaboration in research with scientists located at other institutions required occasional publication. In researching the grassland scientists, I never found this criterion inadequate. The only scientist of importance who was not included in the collective bibliography was Roscoe Pound, who had published only two titles. Of course, Pound was important! But Pound left active scientific work after 1898, when he began teaching at the University of Nebraska law department; for even this extraordinarily energetic man, two careers ultimately were too much. Therefore, the criterion of three publications did keep out of the collective biography a scientist who had no current participation in the normal science period of the specialty.

The decision to weigh equally sole authorship and coauthorship in counting the number of titles per scientist is based on the work of the

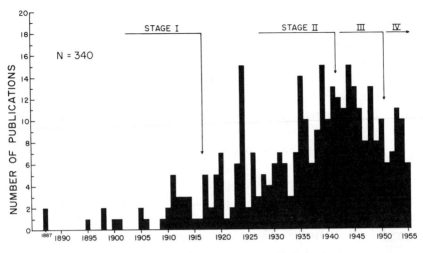

App. fig. 4. Absolute frequency of coauthored publications

Coles in which questionnaire surveys established that contemporary scientists do not discriminate between coauthors in citations of articles.[3]

I collected information on the *professional* careers of the scientists. Several considerations supported this restriction of attention. In the research process of testing the Kuhn-Crane model of specialty growth, I first determined whether the collective biography, confined to professional career information, would meet the model's predictions. If the patterns did not, then I would expect the nonprofessional aspects of the scientists' lives must also be influencing their behavior. Such nonprofessional factors might include father's occupational status, sexual identification, place of residence, and military service, to choose variables that have been investigated by historians of other groups of scientists. As it turned out, professional variables explained the scientists' behavior and met the conditions of the Kuhn-Crane model. I consider the capacity of the professional variables to explain the professional pattern as oblique confirmation of the Kuhnian expectation that normal science is insulated from the distal social background. This confirmation does not exclude the possibility that social background variables may have contributed, but it does imply that they exercised their influence through the profession, rather than by disrupting it.[4]

Biographical information was entered into printed forms that provided for a total of 155 items of information. While there is no need to reproduce all the items of information here, I can note that they covered each subject's vital statistics (birth, marriage, death), social background (place of birth, ethnicity, religion, father's education, and occupation), education (baccalaureate through highest degree), employment experience (through seven job titles), expeditions, professional societies membership, professional offices held, honors, civil and private (nonprofessional) offices held, and military service. In the compilation of employment statistics, every change in the title of employment, regardless whether the employer was also changed, was counted as a new job; hence, the jump from assistant professor to associate professor within the same university was counted as a jump from one job to a second job. It was not always possible to obtain complete information on or to be certain of employment, year to year, for each of the fifty-eight scientists. Occasionally, I obtained information from nonsystematic sources, such as identifications on the author by-lines of articles, but I did not use these if I could not be certain of dates involved. Absolutely complete information on all subjects would of course somewhat alter the shape of text figs. 5, 6, and 7, but I am confident that the general pattern would not be changed and that the generalizations I have drawn from the data would stand.

The fifty-eight scientists who published three or more articles are listed in Appendix table 3.

The information on the lives of these fifty-eight scientists was obtained primarily from the various editions of *American Men of Science*. Information was also obtained from the archives of the University of Nebraska, Lincoln, and the Alumni Association Office of the University of Nebraska, Lincoln. Information from the Alumni Association Office has been used with statistical anonymity for the subjects concerned. Some information concerning the employment history of the subjects was garnered from institutional affiliations noted in articles and from nonsystematic sources, such as the *International Addressbook of Botanists* (1931).

Citation Analysis

Citation analysis is now a prominent feature of science studies. Drawing on the vast, computerized data of the *Science Citation Index*, researchers are able to chart the flow of recent scientific work from one specialty to another. Analytical tools, such as co-citation coupling, which follows the joint appearances of authors in bibliographies, enable sociologists and science policy researchers to monitor the emergence, blossoming, and decay of microspecialties. It had been hoped at one time, perhaps vainly, that citation analysis would be used by historians to study the shift of scientific research from one intellectual orientation to another. So Thomas Kuhn wrote in 1962:

> If I am right that each scientific revolution alters the historical perspective of the community that experiences it, then that change of perspective should affect the structure of postrevolutionary textbooks and research publications. One such effect—a shift in the distribution of the technical literature cited in the footnotes to research efforts— ought to be studied as a possible index to the occurrence of revolutions.[5]

After nearly two decades, only one historical citation study—and this by a sociologist!—has emerged to meet Kuhn's challenge.[6]

Several obstacles may have blunted historians' enthusiasm for citation research. First is the considerable labor of collecting the data and putting them into usable form. For topics that are already well known, such as the chemical revolution associated with Lavoisier or the Darwinian revolution, historians could be suspicious that the citation study would turn up anything fundamentally new. For a topic the history of which is not well known, such as the present study on ecology, historians might well assume that their labor is better spent plotting the narrative and intellectual development, rather than playing with citations; the former certainly would result in fundamental new understanding, and the latter would only be comprehensible in light of a narrative. Another obstacle has been that Kuhn's suggestions about citation shifts during revolutions were so brief that they did not suggest an unambiguous research program. The efforts of H. Gilman McCann to chart the shift of citations in the chemical revolution, for instance, have shown that the citations to the oxygen hypothesis increased over time and citations to phlogiston decreased, but he does not explain how this result verifies a Kuhnian theory of noncumulative progress as distinguished from the cumulative theory of scientific progress of the positivists.[7]

Despite the difficulties in using citation data to decide between competing theories of scientific progress, citation data do lend themselves to one use specified by Kuhn: as an index, if not of revolution, then of change. Citation data conveniently summarize and describe, and I have used them in this way. The story of grassland ecology over two generations is potentially very long. Exhaustive rendition of that history of fifty-eight scientists, of their intellectual development, of their sociological arrangements, and analysis of this history in terms of current theories of specialty development in the sociology and philosophy of science would require a far longer study than this one already is. It would be a study outrunning even the most generous reader's interest. Consequently, I have turned to citation analysis specifically for summary description.

To conduct the analysis of the historical citations in the specialty, I had to restrict the universe of citations from which I drew my data for study. I suppose I could have entered all the bibliographical references from all plant science publications in the United States over the history of the specialty, or, with less ambition, restricted my data to references from all publications in ecology. Of course, in both cases, the manpower requirements of data collection and preparation would be so great as to make any payoff relatively negligible. I have therefore drawn my citations from a restricted portion of the total bibliography in grasslands ecology of the midwestern United States. The citations were taken from all the articles in the bibliography published by the fifty-eight subject scientists. This restricted bibliography amounted to 282 publications, with 5,717 citations, from 1895 through 1955. Of these 282 publications, 15 were books or review articles with archival bibliographies from which no citations were recorded for study. Another 57 publications contained no bibliographical references or footnotes. In the remaining 210 serial titles, some 67 citations were unidentifiable, carrying, in most cases, neither name or author nor title of the cited publication, for which only a date would be given. The 5,717 citations are the bibliographical references from the 210 publications.

In most instances, the citations were the notes or footnotes in these publications. In a few cases, in which there were no notes or footnotes, the bibliographical entries were accepted. The information taken (for machine records) included the author(s), title, and date of the publication in which the citation appeared, and the author(s), short title, and date of publication of the title to which citation was made. If the subject publication contained no citations, this was also noted.

The restrictions in the bibliography from which I drew the citations limit what hypotheses can be tested with the data. No comparison can be made of citing behavior of sustaining contributors and of marginal contributors to the specialty. We cannot know whether marginal contributors attempted to fit their work, by references, into the microparadigm in the same manner as did sustainors. No comparison can be made between citing behavior in grasslands ecology and that in another specialty, such as forest ecology, in which Clementsian theory was also used. No comparison can be made between levels of biological theorization, as between grasslands specialty and the discipline of botany as a whole. No comparison can be made between the citing behavior of a scientist in grasslands ecology and his citing behavior in any other specialty to which he or she contributed. While these limitations seem severe, we must notice that such comparisons have never been made, not even by researchers using the *Science Citation Index* tapes for current specialties. Beginnings are more humble than the universal knowledge such comparisons require.

What can be studied are the citation patterns from a variety of perspectives, such as the employment of the scientists and stages of growth. If citation patterns shifted over time and if they differed according to these variables at all, then we should be able, in a preliminary way, to relate these shifts to the intellectual stages of the specialty. The shifts will not *prove* that the intellectual history of the specialty passed through the stages I have discovered, and certainly we are not able to *prove* that intellectual history followed the Kuhnian scenario. But the patterns strengthen our confidence in the sociological stages of growth outlined in chapter five and in the intellectual history narrated throughout this study.

I devised several measures of citing behavior: aggregate citation counts, gross rate of citation, gross rate of citation by stage of specialty growth, and most frequently cited titles. Certainly these are not the only measures the citation data are capable of supporting, and certainly these are not the most sophisticated measures, when compared to measures, such as co-citation coupling, employed by scholars in contemporary science studies. My purpose in this study has been to integrate cliometric techniques with the history of science. I have therefore attempted to introduce citation measures close to the natural presentation of footnotes, notes, and bibliographical entries in the scientific literature. I suspect that most readers will find the normalized citation measure of in-print year/manpower year to be somewhat esoteric, since it is not an absolute or natural concept. The in-print year/manpower year (discussed technically below) can be read, to make a contemporary analogy, much as EPA mileage estimates of new automobiles are read. When two different automobiles have EPA highway mileage estimates of twenty miles per gallon and thirty miles per gallon respectively, we understand that these numbers do not refer to the mileages a driver would actually obtain on the highway. The estimates refer to mileages obtained under standardized, controlled conditions and permit comparison of the performances of the automobiles relative to one another. Similarly, when I state that a group of scientists cited a literature at a rate of 0.13 citations/in-print year/manpower year and cited another literature at a rate of 0.08 citations/in-print year/manpower year, the reader should compare these rates to each other, rather than trying to imagine a "typical" scientist actually citing the titles 0.13 or 0.08 times a year.

The "aggregate citation count" is simply the absolute number of times all titles by a scientist were cited during the entire lifetime of the specialty (1895–1955). See Appendix tables 7 and 8 for aggregate citation counts of the fifty-eight subject scientists and the nonsubject scientists. To isolate as many nonsubject scientists as possible, whose works may have been important in the specialty, I have accumulated aggregate citation counts of those nonsubject scientists whose works gathered at least fourteen citations. This conservative criterion is not only below the subject scientists' median value, but below also their modal aggregate citation count. This criterion pulled out nineteen nonsubject scientists.

"Gross rate of citation" may be defined several ways. One possible definition would be the rate of actual use, that is, the aggregate number of citations divided by the number of years between the dates of the first and last citations of any of the scientist's works. Thus, if one scientist's titles were cited 100 times between 1910 and 1935, his gross rate of citation would be 4.0 citations per year. This rate figure is not easy to relate to the definition of aggregate citation counts, however, since "aggregate" means, in the previous context, all years of the specialty's history, not just the years in which citations appeared. Consequently, I have chosen a different rate of citation which takes the entire history of the specialty, subsequent to the scientist's first publication, for the base. (I am not counting the publication of the title itself as the first citation of the title, a practice adopted by D. K. de Solla Price in 1976.) According to this second definition, gross rate of citation is the aggregate number of citations divided by the number of years between the date of publication of the earliest cited work and 1955. For example, consider a scientist with 100 citations to his publications, the earliest of which was published in 1910. In this case, his gross rate of citation would be 100/(1955 − 1910), or 2.2 citations per year. This second definition provides us with a good sense of the use of a scientist's works over this completed history of the specialty (though it might not be useful in contemporary science studies in analyzing citations in specialties that remain active).[8]

I have also sorted out the gross rate of citation according to which stage of the life cycle of grassland ecology the earliest cited title was published. The usefulness of this measure is described in the text.

Defining which cited titles were cited "most frequently" is, of course, arbitrary. My choice of criteria for the definition was guided by my interest in bringing into the analysis as much literature as possible. I did not want to screen out cited titles that grassland scientists might have considered important because of an artificial definition of my own. Therefore, after some preliminary analysis of the data, I defined most frequently cited titles in terms (a) of an aggregate citation count of at least nine, or (b) a gross rate of citation of 0.5 (i.e., cited at least once every two years). Several considerations supported these criteria. Derek Price found for contemporary scientific publications that the mean rate of citation was once a year for ten years.[9] Criterion (a) therefore would capture the publications that historically were cited as often as the average contemporary scientific publication is cited. Toward the end of grassland ecology's life cycle, many publications would not be around for ten years before 1955. They could, nevertheless, have been cited highly during a certain period; for example, an article may have been published in 1948, and not cited until 1951, when it began a citation rate of one citation a year. Was it highly cited? I have no interest in excluding such a title by a preconceived definition. I was interested in letting the data inform me what highly cited meant, so I arbitrarily selected a gross citation rate of 0.5 as an additional criterion of most frequently cited. These two criteria isolated fifty-four titles out of all the titles cited by the fifty-eight grassland scientists.

To compare the citation rates of generations of titles between groups of scientists and between periods, I normalized the citation data with a mathematical concept of citation per in-print year and manpower year. We can see the need for, and character of, this normalization by the following example. Let us suppose we have two groups of ten titles each. Each group received a total of one hundred citations to its titles in a ten-year period. Therefore, an unadjusted gross rate of citation would be ten citations per year for each group. Let us suppose, however, that in one group (A), half of the ten titles comprising the group were in print for all ten years, while the other half were in print for only five years. The titles of the other group (B), meanwhile, were all in print for all ten years. Since the one hundred citations to group A were to articles that had been in print for fewer years than the articles in group B, we would naturally think that, even though the unadjusted gross rate of citation of both groups was equal, the citation rate of group A reflected a more frequent citation of the individual titles. To adjust for this difference, I defined the generational groups of grassland titles in terms of the total number of years the individual titles in each group had been in print for any period. In the hypothetical example above, group A would have a total of seventy-five "in-print" years, while group B would have a total of one hundred "in-print" years. Thus group A would have a citation rate of 1.33 citations/in-print year, while group B would have a citation rate of 1.0 citations/in-print year.

We may still have questions about the example above. What if there are only fifty scientists citing group A, while one hundred scientists cite group B? We would naturally evaluate the citation rate of group A as higher still, since fewer scientists were making the total of one hundred citations to it. Thus, group A's citation rate of 1.33 citations/in-print year would have to be adjusted for the number of scientists in any year citing it. If we assume that all fifty scientists were available all ten years to cite group A titles, then group A citation rate would be 1.33 divided by 500 manpower years (fifty scientists multiplied by the ten years each was available), or 0.003 citations per in-print year per manpower year. Group B would be cited, by this adjustment, 0.001 citations per in-print year per manpower year. Since these normalized citation rates are so small, I have arbitrarily multiplied by one hundred to raise them to a readable level. Of course, this multiplication means that the rate (citations/in-print year/manpower year) cannot be read literally, but only used for comparison with other rates.

The formula used for normalizing the citation rates of grassland titles is, therefore:

$$NCR = [(C/PY) \times 100]/MY$$

Where:

NCR is the normalized citation rate;

C is the absolute number of citations to a literature in a period of years;

PY is the number of years the publications are in print in the period;

MY is the number of years the scientists work in a specified category of employment within the historical period.

Appendix Table 1
Bibliographies of Research Specialties

Specialty	No. titles in bibliography	Dates of bibliography	No. of scientists	Source
Rural sociology	403	1941−66	221	Crane, pp. 173−76
Fine group algebra	305	1934−68	102	Crane, pp. 173−76
Radio astronomy	547	1946−66	n.a.	Edge, Mulkay, pp. 80−81
Grassland ecology	535	1895−1955	372	

Appendix Table 2
Beta Test of Bibliography

Period	B	P	F	df	
1896−1916	0.06437	NS	1.572	1,20	
1917−41	0.53692	0.005	16.479	1,23	significantly different from zero
1942−50	0.30	NS	0.336	1,7	not significantly different from zero
1951−56	-3.25714	0.01	26.202	1,4	significantly negative

Appendix Table 3
The Subject List

Aikman, John M.	Hurtt, Leon C.
Albertson, Fred W.	Keim, Franklin E.
Aldous, Alfred E.	Kramer, Joseph
Allred, Berten W.	McIlvain, Ernest H.
Anderson, Kling L.	Mueller, Irene Marian
Bergman, Herbert F.	Noll, William C.
Bessey, Charles E.	Pool, Raymond J.
Branson, Farrel A.	Riegel, David A.
Braun, Emma Lucy	Robertson, Joseph H.
Bruner, William Ed.	Rydberg, Per Axel
Cain, Stanley A.	Sampson, Arthur W.
Clarke, Sidney E.	Sarvis, Johnson Thatcher
Clements, Edith S. (Mrs. Frederic E.)	Savage, David A.
Clements, Frederic E.	Schaffner, John Henry
Condra, George E.	Sears, Paul Bigelow
Cornelius, Donald R.	Shantz, Homer L.
Costello, David F.	Shelford, Victor E.
Cowles, Henry Chandler	Shimek, Bohumil
Crist, John W.	Stoddard, L.
Darland, Raymond W.	Tomanek, Gerald W.
Dyksterhuis, Edsko Jerry	Transeau, Edgar Nelson
Elias, Maxim K.	Turner, George T.
Flory, Evan L.	Vestal, Arthur G.
Frolik, Anton L.	Visher, Stephen Sargent
Fuller, George D.	Voigt, John Wilbur
Gleason, Henry A.	Weaver, John E.
Hanson, Herbert C.	Whitman, Warren C.
Hayden, Ada	Woolfolk, Edwin J.
Hopkins, Harold H.	Zink, Sarah Ellen

Appendix Table 4
Academic Centers of Grassland Research*

Center	Scientist	Degree origin**	Years of employment
University of Nebraska, Lincoln			
	Bessey	Iowa, 1879	1884–1915
	Clements, F.	Nebraska, 1898	1898–1907
	Condra	Nebraska, 1902	1902–18
	Darland	Nebraska, 1947	1946–48
	Frolik	Wisconsin, 1936	1932–38
	Pool	Nebraska, 1913	1905–48
	Sears	Chicago, 1922	1925–27
	Weaver	Minnesota, 1916	1915–52
	Keim	Cornell, 1927	1914–55

Appendix Table 4—Continued

Center	Scientist	Degree origin**	Years of employment
Kansas State College, Fort Hays			
	Albertson	Nebraska, 1937	1918−51
	Anderson	Nebraska, 1951	1938−55
	Branson	Nebraska, 1952	1947−48
	Hopkins	Nebraska, 1949	1946−55
	Riegel	Ft. Hays, (M.A.) 1939	1939−55
	Tomanek	Nebraska, 1951	1947−55
University of Illinois			
	Gleason	Columbia, 1906	1903−10
	Shantz	Nebraska, 1905	1926−28
	Shelford	Chicago, 1907	1914−46
	Vestal	Chicago, 1915	1929−49 (?)
University of Chicago			
	Cowles	Chicago, 1898	1898−1934 (?)
	Fuller	Chicago, 1913	1909−34
	Shelford	Chicago, 1907	1909−14
University of California, Berkeley			
	Cornelius	Nebraska, 1949	1946−55
	Sampson	George Washington, 1917	1923−36
Iowa State College, Ames			
	Aikman	Nebraska, 1928	1927−51
	Hayden	Iowa, Ames, 1918	1907−38
North Dakota Agricultural College			
	Hanson	Nebraska, 1925	1930−40
	Whitman	Wisconsin, 1939	1938−55
Ohio State University, Columbus			
	Transeau	Michigan, 1904	1915−46
	Schaffner	Michigan, (M.A.) 1894	1897−1918
New York Botanical Garden			
	Rydberg	Columbia, 1898	1899−1927
	Gleason	Columbia, 1906	1919−51

*Defined as employing institutions with at least two scientists, each of whom published at least three articles on grassland ecology.

**Ph.D., unless otherwise specified.

Appendix Table 5
Coauthorships

Scientist	Doctoral institution	No. of coauthored titles
Weaver	Minnesota (teaching at Nebraska)	55
Albertson	Nebraska	13
Clements, F.	Nebraska	11
Hanson	Nebraska	8
Whitman	Wisconsin	6
Darland	Nebraska	6

Appendix Table 6
Key to Authors in Clements-Weaver Complex (fig. 8)

1. F. Larson	23. W. Hansen
2. D. White	24. J. Kramer
3. G. Loder	25. M. Reed
4. R. Petersen	26. R. Pool
5. E. Hegelson	27. F. Jean
6. C. Vorhies	28. J. Robertson
7. M. Morris	29. R. Fowler
8. L. Love	30. I. Mueller
9. E. Clements	31. W. Noll
10. G. Goldsmith	32. L. Stoddard
11. R. Chaney	33. N. Rowland
12. V. Shelford	34. A. Thiel
13. R. Pound	35. S. Shively
14. W. Bruner	36. M. Weldon
15. J. Crist	37. V. Hougen
16. R. Darland	38. H. Biswell
17. T. Fitzpatrick	39. J. Voigt
18. R. Fox	40. F. Branson
19. R. Lipps	41. E. Zink
20. E. Flory	42. G. Tomanek
21. W. Himmel	43. L. Launchbaugh
22. G. Harmon	44. H. Hopkins

Appendix Table 7
Aggregate Citation Counts of Subject Scientists

Rank	Scientist	Aggregate citations
1.	Weaver, J. E.	679
2.	Clements, F. E.	252
3.	Albertson, F.	178
4.	Hanson, H. C.	85
5.	Savage, D. A.	84
6.	Shantz, H. L.	76
7.	Sampson, H. C.	71

Appendix Table 7—Continued

Rank	Scientist	Aggregate citations
8.	Visher, S. S.	69
9.	Shimek, B.	63
10.	Gleason, H. A.	58
11.	Condra, G. E.	42
12.	Aldous, A. E.	40
13.	Sarvis, J. T.	37
14.	Stoddard, L.	37
15.	Costello, D. F.	35
16.	Cowles, H. C.	32
17.	Dyksterhuis, E. J.	27
18.	Pool, R. J.	27
19.	Transeau, E. N.	27
20.	Darland, R. W.	26
21.	Noll, W. C.	25
22.	Bruner, W. E.	23
23.	Bessey, C. E.	22
24.	Flory, E. L.	20
25.	Mueller, I.	19
26.	Riegel, D. A.	19
27.	Sears, P.	19
28.	Robertson, J. H.	18
29.	Rydberg, P. A.	18
30.	Vestal, A. G.	18.
31.	Crist, J.	17
32.	Fuller, G. D.	17
33.	Clarke, S. E.	16
34.	Hopkins, H. H.	16
35.	Whitman, W.	16
36.	Aikman, J. M.	15
37.	Allred, B. W.	15
38.	McIlvain, E. H.	15
39.	Shelford, V. E.	15
40.	Anderson, K. L.	13
41.	Cain, S.	13
42.	Cornelius, D. R.	13
43.	Elias, M.	12
44.	Schaffner, J. H.	12
45.	Keim, F. D.	10
46.	Kramer, J.	10
47.	Frolik, A. L.	9
48.	Tomanek, G. W.	9
49.	Turner, G. T.	8
50.	Voigt, J. M.	8
51.	Woolfolk, E. J.	7
52.	Zink, S. E.	7
53.	Braun, E. L.	6
54.	Clements, E.	6
55.	Hurtt, L. C.	5
56.	Branson, F. A.	4
57.	Hayden, A.	3
58.	Bergman, H. F.	1

Appendix Table 8
Aggregate Citation Counts of Nonsubject Scientists

Rank	Scientist	Aggregate citations
1.	Fitzpatrick, T.	61
2.	Hansen, W.	43
3.	Pound, R.	33
4.	Livingston, B.	29
5.	Hitchcock, A.	27
6.	Russel, J.	26
7.	Smith, J.	26
8.	Love, L.	24
9.	Briggs, L.	23
10.	Campbell, R.	19
11.	Stewart, G.	19
12.	Alway, F.	17
13.	Duley, F.	17
14.	Bennett, H.	16
15.	Harshberger, J.	16
16.	Gray, A.	15
17.	Tansley, A. G.	15
18.	Todd, J.	15
19.	Forsling, C.	14
20.	Braun-Blanquet	7

Appendix Table 9
Gross Rates of Citation of Subject Scientists

Rank	Scientist	Gross rate of citation	(Rank by aggregate citation; app. table 7)
1.	Weaver, J. E.	16.98	(1)
2.	Albertson, F.	9.37	(3)
3.	Clements, F. E.	4.06	(2)
4.	Savage, D. A.	3.82	(5)
5.	Dyksterhuis, E. J.	2.70	(17)
6.	Darland, R. W.	2.36	(20)
7.	Hanson, H. C.	2.24	(4)
8.	Costello, D. F.	2.19	(15)
9.	McIlvain, E. H.	2.14	(37)
10.	Stoddard, L.	1.85	(14)
11.	Sampson, A. W.	1.65	(6)
12.	Voigt, J. M.	1.60	(50)
13.	Shantz, H. L.	1.55	(7)
14.	Visher, S. S.	1.50	(8)
15.	Mueller, I.	1.36	(25)
16.	Aldous, A. E.	1.29	(12)
17.	Tomanek, G. W.	1.29	(48)
18.	Noll, W. C.	1.25	(21)
19.	Gleason, H.A.	1.14	(10)
20.	Hopkins, H. H.	1.14	(33)
21.	Shimek, B.	1.07	(9)
22.	Sarvis, J. T.	1.06	(13)
23.	Branson, F. A.	1.00	(56)

Appendix Table 9—Continued

Rank	Scientist	Gross rate of citation	(Rank by aggregate citation; app. table 7)
24.	Flory, E. L.	0.95	(24)
25.	Whitman, W.	0.94	(34)
26.	Anderson, K. L.	0.88	(40)
27.	Condra, G. E.	0.86	(11)
28.	Robertson, J. J.	0.82	(28)
29.	Bruner, W. E.	0.82	(22)
30.	Cornelius, D. R.	0.81	(42)
31.	Riegel, D. A.	0.76	(26)
32.	Allred, B. W.	0.75	(36)
33.	Zink, S. E.	0.70	(52)
34.	Clarke, S. E.	0.64	(32)
35.	Pool, R. J.	0.63	(18)
36.	Sears, P.	0.63	(27)
37.	Cowles, H. C.	0.57	(16)
38.	Turner, G. T.	0.57	(49)
39.	Aikman, J. M.	0.52	(35)
40.	Cain, S.	0.52	(41)
41.	Crist, J. W.	0.52	(31)
42.	Transeau, E. N.	0.52	(19)
43.	Vestal, A. G.	0.43	(30)
44.	Fuller, G. D.	0.39	(31)
45.	Woolfolk, E. J.	0.39	(51)
46.	Frolik, A. L.	0.36	(47)
47.	Shelford, V. E.	0.36	(38)
48.	Keim, F. D.	0.34	(45)
49.	Kramer, J.	0.32	(46)
50.	Bessey, C. E.	0.31	(22)
51.	Rydberg, P. A.	0.30	(29)
52.	Elias, M.	0.24	(43)
53.	Schaffner, J. H.	0.21	(44)
54.	Braun, E. L.	0.15	(53)
55.	Hurtt, L. C.	0.13	(55)
56.	Clements, E.	0.12	(54)
57.	Hayden, A.	0.08	(57)
58.	Bergman, H. J.	0.03	(58)

Appendix Table 10
Gross Rates of Citation of Nonsubject Scientists

Rank	Scientist	Gross rate of citation	(Rank by aggregate citation; app. table 8)
1.	Hansen, W.	3.07	(2)
2.	Fitzpatrick, T.	1.09	(1)
3.	Love, L.	0.96	(8)
4.	Stewart, G.	0.79	(11)
5.	Campbell, R.	0.73	(10)
6.	Russell, J.	0.67	(6)
7.	Bennett, H.	0.59	(14)
8.	Livingston, B.	0.56	(4)
9.	Forsling, C.	0.54	(19)
10.	Duley, F.	0.53	(13)
11.	Pound, R.	0.52	(3)
12.	Briggs, L.	0.48	(9)
13.	Hitchcock, A.	0.47	(5)
14.	Smith, J.	0.41	(7)
15.	Alway, F.	0.40	(12)
16.	Harshberger, J.	0.29	(15)
17.	Tansley, A. G.	0.29	(17)
18.	Braun-Blanquet	0.21	(20)
19.	Todd, J.	0.19	(18)
20.	Gray, A.	0.15	(16)

Appendix Table 11
Gross Citation Rates by Stage

(Stage in which first citation to work appeared)			
I. 1895−1916		II. 1917−41	
Scientist	Gross rate	Scientist	Gross rate
1. Weaver	16.98	1. Albertson	9.37
2. Clements, F. E.	4.06	2. Savage	3.82
3. Sampson	1.65	3. Hansen, W.	3.07
4. Shantz	1.55	4. Hanson, H.	2.24
5. Visher	1.50	5. Costello, D.	2.19
6. Gleason	1.14	6. Stoddard	1.85
7. Fitzpatrick	1.09	7. Mueller	1.36
8. Shimek	1.07	8. Aldous	1.29
9. Condra	0.86	9. Noll	1.25
10. Russel	0.67	10. Hopkins	1.14
11. Pool	0.63	11. Sarvis	1.06
12. Cowles	0.57	12. Love	0.96
13. Livingston	0.56	13. Flory	0.95
14. Pound	0.52	14. Whitman	0.94
15. Transeau	0.52	15. Anderson	0.88
16. Briggs	0.48	16. Bruner	0.82
17. Hitchcock	0.47	17. Robertson	0.82
18. Vestal	0.43	18. Cornelius	0.81
19. Smith, J.	0.41	19. Stewart	0.79
20. Alway, F.	0.40	20. Riegel	0.76
21. Fuller	0.39	21. Allred	0.75
22. Shelford	0.36	22. Campbell	0.73
23. Bessey	0.31	23. Clarke	0.64
24. Rydberg	0.30	24. Sears	0.63
25. Harshberger	0.29	25. Bennett	0.59
26. Tansley	0.29	26. Turner	0.57
27. Elias	0.24	27. Forsling	0.54
28. Schaffner	0.21	28. Duley	0.53
29. Todd	0.19	29. Aikman	0.52
30. Braun	0.15	30. Cain	0.52
31. Gray	0.15	31. Crist	0.52
32. Clements, E.	0.12	32. Woolfolk	0.39
33. Bergman	0.03	33. Frolik	0.36
		34. Keim	0.34
		35. Kramer	0.32
		36. Hurtt	0.13
		37. Hayden	0.08

III. 1942−50		IV. 1951−55	
Scientist	Gross rate	Scientist	Gross rate
1. Dyksterhuis	2.70	1. Branson	1.00
2. Darland	2.36		
3. McIlvain	2.14		
4. Voigt	1.60		
5. Tomanek	1.29		
6. Zink	0.70		

Appendix Table 12
The Founding Literature
Generations of Grassland Literature
(In order of publication)

1. Eugenius Warming, *Lehrbuch der ökologischen Pflanzengeographie; eine Einführung in die Kenntnis der Pflanzenvereine*, translated by Dr. Emil Knoblaugh (Berlin: Gebruder Borntraeger, 1896), and all subsequent translations.

2. A. F. W. Schimper, *Pflanzen-geographie auf physiologischer grundlage* (Jena: G. Fischer, 1898).

3. Roscoe Pound and Frederic E. Clements, *The Phytogeography of Nebraska*, second edition revised (Lincoln, Nebraska: Published by the Seminar, 1900; the first edition of 1898, which was the thesis version, was destroyed in a fire).

4. H. C. Cowles, "The Physiographic Ecology of Chicago and Vicinity; A Study of the Origin, Development, and Classification of Plant Societies," *Botanical Gazette* 31 (1901): 73–108.

5. J. J. Thornber, "The Prairie-Grass Formation in Region I," Botanical Survey, University of Nebraska *Report* 5 (1901).

6. Frederic E. Clements, *Research Methods in Ecology* (Lincoln, Neb.: University Publishing Co., 1905).

7. E. N. Transeau, "Forest Centers of Eastern America," *American Naturalist* 39 (1905): 875–89.

8. L. H. Harvey, "Floral Succession in the Prairie-Grass Formation of Southeastern South Dakota," *Botanical Gazette* 46 (1908):81–108, 277–98.

9. B. Shimek, "The Prairies," Laboratory of Natural History, University of Iowa *Bulletin* 6 (1911):169–240.

10. R. J. Pool, "A Study of the Vegetation of the Sandhills of Nebraska," *Minnesota Botanical Studies* 4 (1914):189–312.

11. Frederic E. Clements, *Plant Succession: An Analysis of the Development of Vegetation* (Washington, D.C.: Carnegie Institution, 1916).

Appendix Table 13
Normal Science Literature A
Generations of Grassland Literature
(In order of publicaton)

1. H. L. Shantz, "Natural Vegetation as an Indicator of the Capabilities of Land for Crop Production in the Great Plains Area," U.S. Department of Agriculture, Bureau of Plant Industry, *Bulletin* 201 (1911). (Shantz's paper is included under this generation, since it received its highest citation rate after 1928, as did all other normal science titles.)

2. F. E. Clements, *Plant Succession: An Analysis of the Development of Vegetation* (Washington, D.C.: Carnegie Institution, 1916).

3. J. E. Weaver, *The Ecological Relations of Roots* (Washington, D.C.: Carnegie Institution, 1919).

4. A. W. Sampson, "Plant Succession in Relation to Range Management," U.S. Department of Agriculture, *Bulletin* 791 (1919).

5. J. T. Sarvis, "Composition and Density of the Native Vegetation in the Vicinity of the Northern Great Plains Field Station," *Journal of Agricultural Research* 19 (1920):63–72.

6. F. E. Clements, *Plant Indicators* (Washington, D.C.: Carnegie Institution, 1920).

7. J. E. Weaver, *Root Development in the Grassland Formation* (Washington, D.C.: Carnegie Institution, 1920).

8. J. T. Sarvis, "Effects of Different Systems and Intensities of Grazing upon the Native Vegetation at the Northern Great Plains Field Station," U.S. Department of Agriculture, *Bulletin* 1170 (1923).

9. F. E. Clements and J. E. Weaver, *Experimental Vegetation* (Washington, D.C.: Carnegie Institution, 1924).

Appendix Table 14
Normal Science Literature B
Generations of Grassland Literature
(In order of publication)

1. J. E. Weaver and F. E. Clements, *Plant Ecology* (New York: McGraw-Hill, 1929). Second edition, 1938.

2. A. E. Aldous, "Effect of Different Clipping Treatments on the Yield and Vigor of Prairie Grass Vegetation," *Ecology* 11 (1930):752–59.

3. T. L. Steiger, "Structure of Prairie Vegetation," *Ecology* 11 (1930):170–217.

4. H. Hanson, L. D. Love, and M. S. Morris, "Effects of Different Systems of Grazing by Cattle Upon a Western Wheat-grass Type of Range," Colorado Experiment Station, *Bulletin* 377 (1931).

5. J. E. Weaver and W. J. Himmel, "The Environment of the Prairie," University of Nebraska Conservation and Survey Division, *Bulletin* 5 (1931).

6. J. E. Weaver and T. J. Fitzpatrick, "Ecology and Relative Importance of the Dominants of Tall-grass Prairie," *Botanical Gazette* 93 (1932):113–50.

7. A. E. Aldous, "The Effect of Burning on Kansas Bluestem Pastures," Kansas Agricultural Experiment Station, *Technical Bulletin* 38 (1934).

8. J. E. Weaver and T. J. Fitzpatrick, "The Prairie," *Ecological Monographs* 4 (1934):109–295.

9. F. E. Clements, "The Relict Method in Dynamic Ecology," *Journal of Ecology* 22 (1934): 39–68.

10. J. E. Weaver and G. W. Harmon, "Quantity of Living Plant Materials in Prairie Soils in Relation to Runoff and Soil Erosion," University of Nebraska Conservation and Survey Division, *Bulletin* 8 (1935).

11. J. E. Weaver, V. H. Hougen, and M. D. Weldon, "Relation of Root Distribution to Organic Matter in Prairie Soil," *Botanical Gazette* 96 (1935):389–420.

12. F. W. Albertson, "Ecology of Mixed Prairie in West Central Kansas," *Ecological Monographs* 7 (1937):481–547.

13. D. A. Savage, "Grass Culture and Improvement in the Central and Southern Great Plains," U.S. Department of Agriculture, *Circular* 491 (1939).

14. S. B. Shively and J. E. Weaver, "Amount of Underground Plant Materials in Different Grassland Climates," University of Nebraska Conservation and Survey, *Bulletin* 21 (1939).

15. J. E. Weaver and V. H. Hougen, "Effect of Clipping on Plant Production in Prairie and Pasture," *American Midland Naturalist* 21 (1939):396–414.

Appendix Table 15
Drought Literature
Generations of Grassland Literature
(In order of publication)

1. J. E. Weaver, L. A. Stoddart, and W. Noll, "Response of the Prairie to the Great Drought of 1934," *Ecology* 16 (1935):612–29.

2. J. E. Weaver and F. W. Albertson, "Effects of the Great Drought on the Prairies of Iowa, Nebraska, and Kansas," *Ecology* 17 (1936):567–639.

3. D. A. Savage, "Drought Survival of Native Grass Species in the Central and Southern Great Plains," U.S. Department of Agriculture, *Technical Bulletin* 549 (1937).

4. J. H. Robertson, "A Quantitative Study of True-Prairie Vegetation after Three Years of Extreme Drought," *Ecological Monographs* 9 (1939):433–92.

5. J. E. Weaver and F. W. Albertson, "Major Changes in Grassland as a Result of Continued Drought," *Botanical Gazette* 100 (1939):576–91.

6. J. E. Weaver and F. W. Albertson, "Deterioration of Grassland from Stability to Denudation with Decrease in Soil Moisture," *Botanical Gazette* 101 (1940):598–624.

7. J. E. Weaver and F. W. Albertson, "Deterioration of Midwestern Ranges," *Ecology* 21 (1940):216–36.

8. J. E. Weaver and W. W. Hansen, "Native Midwestern Pastures: Their Origin, Composition, and Degeneration," University of Nebraska Conservation and Survey, *Bulletin* 22 (1941).

9. J. E. Weaver and W. W. Hansen, "Regeneration of Native Midwestern Pastures under Protection," University of Nebraska Conservation and Survey, *Bulletin* 23 (1941).

10. F. W. Albertson and J. E. Weaver, "History of the Native Vegetation of Western Kansas during Seven Years of Continuous Drought," *Ecological Monographs* 12 (1942):23–51.

11. J. E. Weaver, "Competition of Western Wheat Grass with Relict Vegetation of Prairie," *American Journal of Botany* 29 (1942):366–72.

12. J. E. Weaver and F. W. Albertson, "Resurvey of Grasses, Forbs, and Underground Plant Parts at the End of the Great Drought," *Ecological Monographs* 13 (1943):63–117.

13. J. E. Weaver and F. W. Albertson, "Nature and Degree of Recovery of Grassland from the Drought of 1933 to 1940," *Ecological Monographs* 14 (1944):393–479.

Appendix Table 16
Range Management Literature
Generations of Grassland Literature
(In order of publication)

1. L. A. Stoddard and A. D. Smith, *Range Management* (New York: McGraw-Hill, 1943).

2. G. W. Tomanek, "Pasture Types of Western Kansas in Relation to the Intensity of Utilization in Past Years," Kansas Academy of Sciences, *Transactions* 51 (1948):171–96.

3. J. E. Weaver and J. W. Voigt, "Monolith Method of Root Sampling in Studies on Succession and Degeneration," *Botanical Gazette* 111 (1950):286–99.

4. J. W. Voigt and J. E. Weaver, "Range Condition Classes of Native Midwestern Pasture: An Ecological Analysis," *Ecological Monographs* 21 (1951):39–60.

5. H. H. Hopkins, "Ecology of the Native Vegetation of the Loess Hills in Central Nebraska," *Ecological Monographs* 21 (1951):125–47.

Appendix Table 17
Citation Rates of Grassland Literatures (Summary)
(Citations/In-print Year/Manpower Year)

Employer	1895–1916	1917–28	1916–28	1929–49			1935–46	1947–55		1950–55		
	Founding Literature	Founding Literature	Normal Science A	Founding Literature	Normal Science A	Normal Science B	Drought Literature	Drought Literature	Range Management	Founding Literature	Normal Science A	Normal Science B
Academic	0.13	0.09	0.10	0.03	0.11	0.13	0.55	0.24	0.26	0.02	0.32	0.32
Federal	0.10	0.14	0.29	0.00	0.08	0.08	0.11	0.03	0.23	0.00	0.03	0.14
Other	0.03	0.09	0.05	0.02	0.08	0.03	0.00	0.00	0.00	0.00	0.00	0.00

Appendix Table 18
Movement of the Research Front
Grasslands Bibliography

Years	Percent of Citations with Lag of						Median lag (yrs.)
	0 yrs.	1 yr.	2 yrs.	3 yrs.	4 yrs.	5 yrs.	
1895–1899	0.0	4.7	2.3	7.0	7.0	0.0	54
1900–1904	0.0	0.0	(66.7)	(33.3)	0.0	0.0	2.5
1905–09	(5.7)	(11.3)	(13.2)	(9.4)	(15.1)	1.9	4
1910–14	1.9	5.3	6.7	6.9	5.1	5.1	10
1915–19	0.9	2.7	4.2	5.4	5.7	(8.1)	9
1920–24	(2.5)	1.9	5.9	5.0	5.3	4.3	11
1925–29	(3.1)	(8.0)	5.3	6.2	4.9	2.7	11
1930–34	(3.7)	(7.1)	(11.8)	(9.3)	6.9	5.7	6
1935–39	(2.7)	(9.0)	(9.5)	(8.9)	(8.9)	6.0	6
1940–44	2.0	5.7	(9.5)	7.9	(8.8)	(7.1)	6
1945–49	1.4	5.8	3.9	5.5	5.4	5.3	10
1950–54	1.0	4.0	(8.7)	(8.5)	6.8	6.3	7
all years	2.0	6.0	8.0	8.0	7.0	6.0	

Interpretation:

Percentages within parentheses are above the all-years percentage. The research front begins building in 1920; is established between 1930 and 1944; has collapsed after 1944.

NOTES

Introduction

    1. For general introductions to the history of ecology, see Donald Worster, *Nature's Economy: The Roots of Ecology* (San Francisco: Sierra Club Books, 1977); Frank N. Egerton, "A Bibliographical Guide to the History of General Ecology," *History of Science* 15 (1977):189–215; Frank N. Egerton, "Ecological Studies and Observations Before 1900," pp. 311–51, in Benjamin J. Taylor and Thurman J. White, *Issues and Ideas in America* (Norman: University of Oklahoma Press, 1976); Robert P. McIntosh, "Ecology Since 1900," pp. 353–72, ibid.; Robert P. McIntosh, "25 Years of Botany," *Annals of the Missouri Botanical Garden* 61 (1974): 132–65. See also the bibliographical notes to Ronald Tobey, "Theoretical Science and Technology in American Ecology," *Technology and Culture* 17 (1967):7 18–19, and the essays in Frank N. Egerton, et al., *History of American Ecology* (New York: Arno Press, 1977).

    2. Paul B. Sears, *Lands Beyond the Forest* (Englewood Cliffs, N.J.: Prentice-Hall, 1969), p. 4.

    3. David F. Costello, *The Praire World* (New York: Thomas Y. Crowell, 1969), p. 1.

    4. John E. Weaver, "North American Prairie," *American Scholar* 13 (1944):334, 339.

    5. John E. Weaver, *North American Prairie* (Lincoln, Nebr.: Johnsen, 1954), p. viii.

    6. Frederic E. Clements and John E. Weaver, *Experimental Vegetation* (Washington, D.C.: Carnegie Institution, 1924), p. 5.

    7. Laurence Veysey, "Intellectual History and the New Social History," pp. 3–26, in John Higham and Paul K. Conkin, eds., *New Directions in American Intellectual History* (Baltimore and London: Johns Hopkins University Press, 1979).

1. Missionaries for Botany

    1. Title quotation from Luke Lawson to Charles E. Bessey, April 24, 1896, Charles Edwin Bessey Papers, University Archives, the University of Nebraska, Lincoln, Nebraska; hereafter, Bessey Papers. Text quotation from Bryon D. Halsted to C. E. Bessey, December 2, 1908, Bessey Papers.

2. Ida Brockman Cornelius to C. E. Bessey, August 2, 1906, Bessey Papers.

3. Biographical studies of Bessey include Thomas R. Walsh, "Charles E. Bessey: Land-Grant Professor" (Ph.D. diss., University of Nebraska, Lincoln, 1972), which draws on two unpublished memoirs for information on Bessey's youth: R. J. Pool, "A Brief Sketch of the Life and Work of a Great Botanist," and Ernest A. Bessey (C. E. Bessey's son), "Notes on Charles Edwin Bessey at Iowa State College of Agriculture (February 1870, until November 1884)" Joseph Ewan, "Bessey, Charles Edwin," in *The Dictionary of Scientific Biography*, ed. Charles Coulton Gillispie (New York, 1970), 102–4; L. J. Pammel, "Dr. Charles Edwin Bessey," *Prominent Men I Have Met* (Ames: Iowa State College, 1925).

Special treatments of Bessey's career include Richard A. Overfield, "Charles E. Bessey: The Impact of the 'New' Botany on American Agriculture, 1880–1910," *Technology and Culture* 16 (April 1975):162–81; Richard A. Overfield, "Trees for the Great Plains: Charles E. Bessey and Forestry," *Journal of Forest History* 23 (January 1979):18–31, Andrew Denny Rodgers, III, *American Botany, 1873–1892: Decades of Transition*, facsimile of the edition of 1944 (New York and London: Hafner, 1968), passim, and *John Merle Coulter, Missionary in Science* (Princeton, N.J.: Princeton University Press, 1944), esp. pp. 54–62, 112–20; Ronald Tobey, "Theoretical Science and Technology in American Ecology," *Technology and Culture* 17 (October, 1976):718–28; Thomas R. Walsh, "Charles E. Bessey and the Transformation of the Industrial College," *Nebraska History* 52 (1971):383–409.

Bessey provides biographical information in response to a questionnaire attached to Howard Edwards to C. E. Bessey, November 24, 1906, Bessey Papers; his religious views were expressed in reply to another questionnaire, C. E. Bessey to W. H. Howard Nash, May 24, 1909, Bessey Papers.

4. For the date of Bessey's arrival in Lincoln, see C. E. Bessey to L. H. Pammel, April 12, 1898, Bessey Papers. Bessey placed his move in November, but he had been in Lincoln in September for an inaugural address as dean. See also Walsh, "Charles E. Bessey and the Transformation of the Industrial College," p. 383, and Robert N. Manley, *Centennial History of the University of Nebraska, I. Frontier University (1869–1919)* (Lincoln: University of Nebraska Press, 1969), pp. 100–110.

5. C. E. Bessey to [Macmillan], September 21, [1882?], Bessey Papers.

6. Record Book of the Sem. Bot., May 24, 1893, p. 30, Records of the Sem. Bot. Club, 1886–1935, University Archives, the University of Nebraska, Lincoln, Nebraska.

7. This self-characterization was made in C. E. Bessey to John H. Schaffner, June 7, 1909, Bessey Papers; C. E. Bessey to George A. Wardlaw, May 31, 1902, Bessey Papers; "Some Personal Remembrances of Dr. Asa Gray [by C. E. Bessey]", Notes for a talk to the Botanical Seminar, University of Nebraska, Lincoln, November 18/19, 1908, boxed under "speeches and articles," Bessey Papers.

8. On the place of Bessey's text in the history of botanical textbooks, see Charles E. Ford, "Botany Texts: A Survey of Their Development in American

Higher Education, 1643 – 1906," *History of Education Quarterly* 4 (1964):59 – 70. See also the appraisal of it by Bessey's contemporary, J. C. Arthur, as "the recognized standard of instruction," in J. C. Arthur, "Development of Vegetable Physiology," *Botanical Gazette* 20 (September 1895):385.

9. R. H. Ward to C. E. Bessey, January 29, 1881, Bessey Papers.

10. N. H. Winchell to C. E. Bessey, August 30, 1880, Bessey Papers.

11. Volney M. Spaulding to C.E. Bessey, August 14, 1880, Bessey Papers.

12. R. N. Prentiss to C. E. Bessey, October 29, 1880, Bessey Papers.

13. Asa Gray to C. E. Bessey, August 12, 1880, Bessey Papers.

14. Rodgers, *Coulter*, p. 51.

15. A. J. Packard, Jr., co-ed., *The American Naturalist*, to C. E. Bessey, August 10, 1880, Bessey Papers.

16. James McKeen Cattell, ed., *American Men of Science* (New York: Science Press, 1910).

17. Ibid.; also see references in n. 3, above.

18. C. W. von Codler, superintendent of Public Instruction, State of Iowa, to C. E. Bessey, appointment letter, January 4, 1881, Bessey Papers. Bessey's involvement in public school education is treated at length in Walsh, "Charles E. Bessey: Land Grant Professor," pp. 147 – 74.

19. See chap. 2, "The Promise of the High School," below.

20. See the letter of charge, Charles S. Palmer, president of the Natural Science Department, National Education Association, to C. E. Bessey, May 22, 1897, Bessey Papers; also, Ralph Tarr to C. E. Bessey, December 12, 1896, C. S. Palmer to C. E. Bessey, April 13, 1896, C. S. Palmer to C. E. Bessey, August 17, 1896, Bessey Papers.

21. Otis W. Caldwell to C. E. Bessey, March 3, 1909, Bessey Papers.

22. For example, see C. E. Bessey to P. M. Hannibal [?], March [?] 14, 188[?]; C. E. Bessey to John O. Taylor, April 26, 1892[?]; Luke Lawson to C. E. Bessey, April 24, 1896; C. E. Bessey to J. M. Stimson, December 11, 1905; C. E. Bessey to T. M. Hodgman, January 30, 1905, Bessey Papers. Walsh discusses Bessey's teaching and paternalism in "Charles E. Bessey: Land-Grant Professor," pp. 132 – 38.

23. Walsh, "Charles E. Bessey: Land-Grant Professor," pp. 190 – 201.

24. Ibid., pp. 205 – 09.

25. Ibid., pp. 213 – 18.

26. The history of the Sem. Bot. has not been recounted before in detail; for sketches see Walsh, "Charles E. Bessey: Land-Grant Professor," pp. 128 – 32, and David Wigdor, *Roscoe Pound, Philosopher of Law* (Westport, Conn. and London: Greenwood Press, 1974), pp. 29 – 30, 50 – 52.

27. Record Book of the Sem. Bot., p. 5.

28. Herbert John Webber, "On the Trail of the Orange: The Autobiography of an Ordinary Man," dictated, 1945 (Biological Sciences Library, University of California, Riverside).

29. Manley, *Centennial History of the University of Nebraska*, p. 91.

30. J. C. Arthur, "Some Botanical Laboratories of the United States," *Botanical Gazette* 10 (December 1885):404.

31. Quoted from Willa Cather, *My Antonia*, by E. K. Brown, *Willa Cather, A Critical Biography*, completed by Leon Edel (New York: Knopf, 1953), p. 62.

32. Webber, "Autobiography," p. 92.

33. Ibid., p. 94.

34. The prank is related in Mildred R. Bennett, *The World of Willa Cather* (Lincoln: University of Nebraska Press, reprint edition, 1961), pp. 190–91. The Sem. Bot. Record Book notes the bust incident, pp. 21–22.

35. Webber, "Autobiography," p. 97.

36. Ibid.

37. Ibid., pp. 80–82.

38. Record Book of the Sem. Bot., p. 3. Reporting the earlier origin of the group, the record book described the formal organization as a decision for "concerted work on the part of the Sems." (p. 5). This language revealed the strain created as each member went on his own way.

39. Ibid., p. 3.

40. Ibid., p. 4.

41. Ibid., p. 5.

42. Ibid., pp. 5–8; Webber, "Autobiography," describes his summer collecting with Bruner, pp. 105–26.

43. Ibid., pp. 6–8.

44. Ibid., p. 9.

45. Ibid., pp. 11, 16; Webber, "Autobiography," p. 137.

46. Wigdor, *Roscoe Pound*, pp. 69–70.

47. Record Book of the Sem. Bot., p. 13.

48. Both Rodgers, *American Botany*, pp. 200–202, and Overfield, "Bessey," make this mistake of limited perspective. The name "Bessey system" referring to the instructional setting should be distinguished from the "Besseyan system," which denotes Bessey's phylogenetic classification system and is not discussed in this book.

49. Webber, "Autobiography," p. 94; Wigdor, *Roscoe Pound*, pp. 73–74.

50. C. E. Bessey, Report to the Regents, December, 1898, Bessey Papers; on the Russian thistle episode, see Tobey, "Theoretical Science and Technology in American Ecology," pp. 721–22.

51. See Per Axel Rydberg, "Flora of the Sandhills of Nebraska," United States National Herbarium, *Contributions*, vol. 3, no. 3.

52. The university's intent to seek a small, specially appropriated sum for publications was expressed in George E. MacLean to C. E. Bessey, December 17, 1898, Bessey Papers.

53. I prepared the following content analysis of Bessey's incoming correspondence for the year 1887, when the Sem. Bot. group was getting under way.

Bessey's Personal Incoming Correspondence, 1887

Content	*(N = 105)*
Requests for information about botany	33%
Plant exchange	27
Plant identification & consultation	8
Personal & family	5
Professional societies	5
Political, conservation	5
Requests for letters of recommendation	4
Teaching	2
Administration	1
Laboratory equipment	1
Miscellaneous	9

54. C. E. Bessey, Report to the Regents, December, 1898, Bessey Papers. The thesis that ecology emerged under Bessey's aegis, presented in these early sections, will also be found, with less definiteness, in Rodgers, *Coulter*, p. 118, and Paul B. Sears, *The Living Landscape* (New York: Basic Books, 1966), p. 78, where it is hinted.

2. The Creative Tension

1. The standard secondary bibliography on this topic includes Arthur A. Ekirch, Jr., *Man and Nature in America* (New York and London: Columbia University Press, 1963); Hans Huth, *Nature and the American: Three Centuries of Changing Attitudes* (Berkeley and Los Angeles: University of California Press, 1957); Roderick Nash, *Wilderness and the American Mind* (New Haven and London: Yale University Press, 1967); Peter Schmidt, *Back to Nature: Arcadian Myth in Urban America* (New York: Oxford University Press, 1969); Henry Nash Smith, *Virgin Land: The American West as Symbol and Myth*, reprint edition (New York: Vintage Books, n.d.; first edition, 1950); R. Richard Wohl, "The 'Country Boy' Myth and Its Place in American Urban Culture: The Nineteenth-Century Contribution," ed. Moses Rischin, *Perspectives in American History*, 3 (1969):77–156. On the Country Life Commission, see William L. Bowers, *The Country Life Movement in America, 1900–1920* (Port Washington, N.Y.: Kennikat Press, 1974), and the original L. H. Bailey, *The Country-Life Movement in the United States* (New York: Macmillan, 1913). An unusual perspective on the American demand for "nature" is provided in Roderick Nash, "The Exporting and Importing of Nature: Nature-Appreciation as a Commodity, 1850–1980," *Perspectives in American History 12* (1979):517–60.

2. Frederic E. Clements and Irving S. Cutter, *A Laboratory Manual of High School Botany* (Lincoln, Nebr.: University Publishing, 1900). On the dispute between Bessey and Coulter, see below.

3. The history of general circulation periodicals is discussed in Frank L. Mott, *A History of American Magazines*, 5 vols. (Cambridge, Mass.: Harvard University Press, 1957–68); the role of scientific journals in research and support of science in the United States has not been adequately treated; see Mott's remarks, 1:438–50, 2:78–81, 84–93, 3:306–45.

4. Schmitt, *Back to Nature*, p. 3.

5. The literary themes of the antiurban literature are surveyed in Schmitt, *Back to Nature*, passim, and analyzed politically for the decade of the 1920s, in Don S. Kirschner, *City and Country: Rural Responses to Urbanization in the 1920s* (Westport, Conn.: Westport Publishing, 1970). The British antiurban, literary ideology is treated in Raymond Williams, *The Country and the City*, reprint edition (St. Albans [England]: Paladin, 1975).

6. Williams, *The Country and the City*, especially chap. 5, "Town and Country," pp. 61–71.

7. I cannot "prove" this thesis, short of a long and tedious monograph, but the *reasonableness* of the thesis is apparent from recent historical studies of small towns. I have provided a basic set of definitions and a schematic of the relationship between social class in towns and cultural activity in Ronald C. Tobey, "How Urbane is the Urbanite? An Historical Model of the Urban Hierarchy and the Social Motivation of Service Classes," *Historical Methods Newsletter* 7 (September 1974): 259–75.

The reality of the public desire for distant vistas in the urban landscape was documented in the 1950s by Kevin Lynch. In a series of interviews in downtown areas of several major cities, Lynch discovered that the open urban vista, such as that across Central Park in New York City and the open waterfront in Boston, and not the architectural city itself, was the most important component of positive images of the city. See Kevin Lynch, *The Image of the City* (Cambridge, Mass.: Technology Press, 1960).

Specifically discussing midwestern American cities, this thesis has been given tangential support by Walter B. Hendrickson, "Science and Culture in the American Middle West," *Isis* 64 (1973):326–40.

For a bibliography on the literature of small towns, see Tobey, "How Urbane is the Urbanite?"

8. Edward A. Krug, *The Shaping of the American High School, 1880–1920* (Madison, Milwaukee, and London: University of Wisconsin Press, 1960).

9. Ibid., pp. 1–17; Edward A. Krug, *The Secondary School Curriculum* (New York: Harper & Brothers, 1960), pp. 24–31; Theodore R. Sizer, *Secondary Schools at the Turn of the Century* (New Haven and London: Yale University Press, 1964), pp. 6–9; on the role of scientists, see ibid., pp. 73–98.

10. John Elbert Stout, *The Development of High School Curricula in the North Central States from 1860 to 1918* (Chicago: University of Chicago Press, 1921), p. 69.

11. Ibid., p. 69, and table 10, p. 72.

12. Ibid., table 10, pp. 71–73.

13. Ibid., table 10, pp. 69–73.

14. Ibid., App. Tables E and H.

15. Ibid., pp. 153–57; J. Y. Bergen, *Elements of Botany* (Boston and London: Ginn, 1897), pp. 210–18, 213, 217; Asa Gray, *Botany for Young People and Common Schools: How Plants Grow, A Simple Introduction to Structural Botany* (New York: Ivison, Blakeman, Taylor, 1858).

16. Sizer, *Secondary Schools*, pp. 33, 73–76. On Eliot's activities and educational philosophy, see ibid., pp. 77–82, and Hugh Hawkins, *Between Harvard and America: The Educational Leadership of Charles W. Eliot* (New York: Oxford University Press, 1972), chap. 8, "Reforming the Lower Schools," pp. 224–62; Krug, *Shaping of the American High School*, p. 86; Stow Persons, *American Gentility* (New York: Columbia University Press, 1973), pp. 179–202; Laurence R. Veysey, *The Emergence of the American University* (Chicago and London: University of Chicago Press, 1965), pp. 86–89.

17. Sizer, *Secondary Schools*, p. 105.

18. Ibid., pp. 218–19.

19. "Report of the Committee of Ten on Secondary School Studies, with the Reports of the Conferences Arranged by the Committee," Sizer, *Secondary Schools*, p. 237.

20. Ibid.

21. See Charles E. Bessey to P. M. Hannibal, March 14, 1892(?), Letterpress Books, 1878–94, Charles Edwin Bessey Papers, University Archives, University of Nebraska, Lincoln, Nebraska, and C. E. Bessey to John O. Taylor, April 26, 1892, ibid.; hereafter, Bessey Papers.

22. Ibid.

23. Charles E. Bessey to James McKeen Cattell, May 12, 1896, Letterpress Books, 1894–98; Charles E. Bessey to T. M. Hodgman, High School Inspector, January 30, 1905, Letterpress Books, 1903–1905, Bessey Papers.

24. Krug, *Shaping of the American High School*, p. 187; see also, Sizer, *Secondary Schools*, p. 47.

25. Stout, *Development of High School Curricula*, pp. 235–42.

26. Krug, *Shaping of the American High School*, pp. 217–18.

27. Ibid., p. 225.

28. Ibid., pp. 368–69.

29. George Santayana, "The Genteel Tradition in American Philosophy," in *Winds of Doctrine and Platonism and the Spiritual Life*, reprint edition (New York: Harper & Brothers, Harper Torchbooks, 1957), pp. 187–215. The bibliography on high culture and the arts in the United States between the Gilded Age and World War I is enormous and no purpose is served by citing all titles that relate to the genteel tradition. I have found the following general summaries useful guides: Howard Mumford Jones, *The Age of Energy, Varieties of American Experience, 1865–1915*, reprint edition (New York: Viking Press, Viking Compass Edition, 1973), chap. 6, "The Genteel Tradition," pp. 216–258; Stow Persons, *American Gentility*, John Tomisch, *A Genteel Endeavor: American Culture and Politics in the Gilded Age* (Stanford, Calif.: Stanford University Press, 1971). Helen Lefkowitz Horowitz has usefully brought together the philanthropic activities of Chicago's genteel patrons in *Culture and the City, Cultural Philanthropy in Chicago from the 1880s to 1917* (Lexington: University of Kentucky Press, 1976). Donald Meyer explores the ethical basis of the

genteel tradition in textbooks of moral philosophy; D.H. Meyer, *The Instructed Conscience: The Shaping of American National Ethic* (Philadelphia: University of Pennsylvania Press, 1972).

30. Jones, *Age of Energy*, p. 216; Tomisch, *Genteel Endeavor*, pp. 1–26, passim.

31. Persons, *American Gentility*, p. 278.

32. Edward Lurie, *Louis Agassiz, A Life in Science*, abridged reprint edition (Chicago and London: University of Chicago Press, Phoenix Books, 1960), pp. 84–88, where Lurie sketches Agassiz's belief in unity of nature, pp. 133–41, and p. 203. On Joseph Le Conte, see Paul F. Boller, Jr., *American Thought in Transition: The Impact of Evolutionary Naturalism, 1865–1900* (Chicago : Rand McNally, 1969), pp. 15–16.

33. Howard S. Miller, *Dollars for Research: Science and Its Patrons in Nineteenth-Century America* (Seattle and London: University of Washington Press, 1970), chap. 2, "The Works of Creation," pp. 24–47, and chap. 4, "Science and the Community Interest," pp. 71–97, passim.

34. Sociologists, more often than historians, have investigated the impact of the laboratory setting on differentiation of scientific roles, professionalization, and instruction. See the discussion by Joseph Ben-David, *The Scientist's Role in Society: A Comparative Study* (Englewood Cliffs, N.J.: Prentice-Hall, 1971), chap. 7, "German Scientific Hegemony and the Emergence of Organized Science," pp. 108–38, esp. pp. 123–25.

35. For surveys of antebellum science, see George H. Daniels, *American Science in the Age of Jackson* (New York and London: Columbia University Press, 1968); A. Hunter Dupree, *Science in the Federal Government: a History of Policies and Activities to 1940*, reprint edition (New York and Evanston: Harper & Row, Harper Torchbooks/The Science Library, 1957), chaps. 1–6; Stanley M. Guralnick, *Science and the Ante-bellum American College*, Memoirs of the American Philosophical Society, 109 (Philadelphia: American Philosophical Society, 1975); Sally Gregory Kohlstedt, *The Formation of the American Scientific Community: The American Association for the Advancement of Science, 1848–1860* (Urbana: University of Illinois Press, 1976). George H. Daniel's much criticized *Science in American Society, A Social History* (New York: Knopf, 1971), chaps. 6–9, contains useful information, as does the antiquated work, Dirk J. Struik, *Yankee Science in the Making*, new revised reprint edition (New York: Collier Books, 1962). The biographies of Louis Agassiz and Asa Gray also illuminate the period from the perspective of two of its outstanding scientists, both at Harvard; Lurie *Louis Agassiz*, and A. Hunter Dupree, *Asa Gray, 1810–1888, reprint edition (New York, Atheneum, 1968)*.

For general discussions on the roles of two great universities identified with scientific research after the Civil War, Johns Hopkins and Wisconsin, see Frederick Rudolph, *The American College and University, A History* (New York: Knopf, 1962), chaps. 12 and 13, pp. 241–86, and Laurence Veysey, *The Emergence of the American University*, chap. 2, "Utility," and chap. 3, "Research."

For discussions of the New Botany, consult Dupree, *Asa Gray*, Andrew Denny Rodgers, III, *American Botany, 1873–1892: Decades of Transition*, facsimile of the edition of 1944 (New York and London: Hafner, 1968), and *John Merle Coulter*,

Missionary in Science (Princeton, N.J.: Princeton University Press, 1944). Hamilton Cravens's recent work provides a good overview of the "New Biology;" see Hamilton Cravens, *The Triumph of Evolution: American Scientists and the Heredity-Environment Controversy, 1900—1941* (Philadephia: University of Pennsylania Press, 1978), pp. 15—55.

36. For this discussion of specialization, I am drawing on Ben-David, *The Social Role of Scientists*, pp. 142—43, 147—55, and the discussions in this work, chaps. 1 and 5.

37. C. E. Bessey to J. M. Coulter, April 19, 1899, Letterpress Books, 1898—1901, Bessey Papers. Italics in original.

38. J. M. Coulter to C. E. Bessey, April 21, 1899, Bessey Papers.

39. Bessey to Coulter, April 19, 1899, Bessey Papers.

40. C. E. Bessey to A. F. Woods, Bureau of Plant Industry, Department of Agriculture, June 24, 1903, Letterpress Books, 1902—1903, Bessey Papers.

41. C. E. Bessey to A. G. Tansley, December 15, 1905 (1906?), Box 1905 (G)—1906 (Ba), Bessey Papers.

42. H. C. Cowles, Review of F. C. Clements, *Research Methods in Ecology*, *Botanical Gazette* 40 (November, 1905):381—82.

43. Rodgers, *Coulter*, p. 44.

44. Charles E. Bessey, *Botany for High Schools and Colleges* (New York: Henry Holt, 1880), p. iii.

45. Ibid., p. 1.

46. Charles E. Bessey, *The Essentials of Botany*, fifth edition (New York: Henry Holt, 1893), p. xi.

47. Bessey, *Botany*, pp. 3—4.

48. Ibid., pp. 5—14.

49. Ibid., p. iv.

50. Julius von Sachs, *Text-book of Botany, Morphological and Physiological*, trans. Alfred W. Bennett (Oxford [England]: Clarendon Press, 1875), pp. 128—29.

51. Sachs, *Text-book*, trans. Bennett, p. 1. This principle of *naturphilosophie* is discussed in chap. 4, below. For a summary of reductive materialism in German biology before Sachs, see William Coleman, *Biology in the Nineteenth Century: Problems of Form, Function, and Transformation* (New York: John Wiley, 1971), pp. 150—54.

52. Sachs, *Text-book*, trans. Bennett, p. 129.

53. Ibid.

54. Ibid., pp. 130—31.

55. Bessey, *Botany*, pp. 133—34.

56. Sachs, *Text-book*, trans. Bennett, p. 129.

57. Bessey, *Essentials of Botany*, p. ix.

58. The standard work on the new learning in the United States is Morton White, *Social Thought in America; The Revolt Against Formalism*, reprint edition (Boston: Beacon Press, 1957).

59. Charles Saunders Peirce, "Concerning the Author,"reprinted in Justus Buchler, editor, *Philosophical Writings of Peirce*, reprint edition (New York: Dover, 1955), p. 1.

60. Charles Saunders Peirce, "How to Make Our Ideas Clear," reprinted in Buchler, ed., ibid., p. 30.

61. Ibid., p. 31.

3. Quadrat

1. Edouard Rübel, "Über die Entwicklung der Gesellschaftsmorphologie," *Journal of Ecology* 8 (1920):29.

2. The history of modern plant ecology remains without major monographic treatment. While outlines of the major contributions to phytogeography appeared occasionally as précis in nineteenth-century scientific publications, no one has yet constructed a coherent narrative of the field. In addition, there is a wide variety of connections between the scientific specialty and the revolution in European landscape. The rapid growth of urban and suburban gardening, the prolific importation of nonindigenous plant species, the alteration of the landscape by industrialization, and the agricultural revolution powerfully affected the understanding of the relationship between plant and environment and between changing plant distribution and changing geography. We need, in a word, a monograph that does for the nineteenth century what Clarence Glacken's *Traces on the Rhodian Shore* does for the earlier centuries.

Basic scientific documents in the history of plant ecology have been reprinted in Frank N. Egerton, ed., *History of Ecology*, an Arno Press Collection, 52 vols. (New York: Arno Press, 1977). Humboldt's and Aimé Bonpland's *Géographie des plantes* (1807) is reprinted as a separate volume in this collection. Augustin de Candolle, "Geographie botanique" (1820), is reprinted in the volume, Egerton, ed., *Ecological Phytogeography in the Nineteenth Century* (New York: Arno Press, 1977).

On related materials, see Gareth Nelson, "From Candolle to Croisat: Comments on the History of Biogeography," *Journal of the History of Biology* 11 (1978):269–305.

3. I have been unable to locate a general history of isolines in climatic cartography. See the remarks, W. Köppen and R. Geiger, eds., *Handbuch der Klimatologie*, 5 vols. (Berlin: Gebruder Borntraeger, 1936), 1:121. Frank Egerton has reprinted Humboldt's essay, "On Isothermal Lines, and the Distribution of Heat Over the Globe" (1820), in Egerton, ed., *Ecological Phytogeography in the Nineteenth Century*.

4. Alexander von Humboldt and Aimé Bonpland, *Essai sur la géographie des plantes* (Paris: Libraire Schoell, 1807) and Alexander von Humboldt, *Aspects of Nature*, trans. (from the German) Mrs. Sabine, third edition, first English edition, 2 vols. (London: Longman, Brown, Green, and Longmans, 1849).

5. Clements provided an extensive historical background on the quadrat, although he certainly was not familiar with it all in 1897, in Frederic E. Clements, *Plant Succession: An Analysis of the Development of Vegetation* (Washington, D.C.: Carnegie Institution, 1916), pp. 424–25.

6. On the history of the concept of the correlation, see Karl Pearson, "Notes on the History of Correlation," pp. 185–205 in E. S. Pearson and M. G. Kendall, *Studies in the History of Statistics and Probability* (London: Griffin, 1970), vol. 1, and Helen M. Walker, *Studies in the History of Statistical Method* (Baltimore: Williams and Wilkins, 1931), chap. 5, "Correlation," pp. 92–141.

7. John Briquet, "Les méthodes statistiques applicables aux recherches de floristique," *Bulletin de l'herbier bossier* (Geneva) 1 (April 1893):141–58; the study by Hoffmann is *Nachtrage zur Flora des Mittelrheingebietes*, published in *Berichte der oberhess. Gesellschaft fur Natur-und Heilkunde* (Geissen, 1879–87).

8. Oscar Drude, *Deutschlands Pflanzengeographie; Ein geographes Charakterbild der Flora von Deutschland und den angrenzenden Alpen-Sowie Karpathenlaudern* (Stuttgart: Verlag von J. Engelhorn, 1896).

9. Briquet, "Les méthodes statistiques," p. 158. Briquet did not discuss the work of Hult and other botanists of the Uppsala School in the 1880s and 1890s; for a treatment of the school, see Rubel, "Uber die entwicklung der Gesellschafts-Morphologie." A useful review of the development of floristics is provided by David W. Shimwell, *The Description and Classification of Vegetation* (Seattle: University of Washington Press, 1971), esp. chap. 6, "Floristic Systems of Vegetation Description," pp. 185–220. Shimwell provides an epistemological analysis of classification schemes similar to mine, ibid., pp. 42–44. I am indebted to Frank Egerton for a suggestion to examine the relation of the Uppsala School to the points I am making in this chapter.

10. Drude, *Deutschlands Pflanzengeographie*, p. 21.

11. Andrew Denny Rodgers III, *John Merle Coulter, Missionary in Science* (Princeton, N.J.: Princeton University Press, 1944), passim.

12. M. E. Meads, "The Range of Variation in Species of Erythronium," *Botanical Gazette* 18 (1893):134–38.

13. Albert Schneider, "Influence of Anaesthetic on Plant Transpiration," *Botanical Gazette* 18 (1893):57–69; H. L. Russell, "The Bacterial Flora of the Atlantic Ocean off Woods Hole, Mass.; A Contribution to the Morphology and Physiology of Marine Bacteria," ibid., pp. 383–95, 411–17, 439–47; S. G. Wright, "Leaf Movements in Cercis Canadensis," ibid. 19 (1894):215–24; D. T. MacDougal, "The Curvature of Roots," ibid. 23 (1897):307–36.

14. Wright, "Leaf Movements," passim.

15. Russell, "Bacterial Flora," passim.

16. Charles B. Davenport, *Statistical Methods, with Special Reference to Biological Variation* (New York: John Wiley, 1899).

17. Karl Pearson, "Mathematical Contributions to the Theory of Evolution, III; Regression, Heredity, and Panmixia," *Philosophical Transactions of the Royal Society of London* 187A (1896):253–318; E. T. Brewster, "A Measure of Variability and the Relation of Individual Variations to Specific Differences," *Proceedings, American Academy of Arts and Sciences*, 32 (1897):268–80.

18. Charles E. Bessey, "Phylogeny and Taxonomy of the angiosperms," *Botanical Gazette* 34 (1897):145–78.

19. H. C. Cowles, "The Physiographic Ecology of Chicago and Vicin-

ity; A Study of the Origin, Development, and Classification of Plant Societies," *Botanical Gazette* 31 (1901):73–108, 145–82.

20. Record Book of the Sem. Bot., p. 18, University Archives, University of Nebraska, Lincoln, Nebraska; hereafter, Record Book. See also the unpublished reminiscence of Pound by Clements, enclosed with letter, F. E. Clements to Sayre, January 17, 1945 (hereafter, unpublished reminiscence), Box 125, Dr. Edith S. Clements and Dr. Frederic Clements Collection, Division of Rare Books and Special Collections, the Library, University of Wyoming, Laramie, Wyoming; hereafter, Clements Collection.

21. Record Book, p. 22.

22. Ibid., p. 58.

23. Roscoe Pound, "Clements," *Ecology* 35 (1954):113.

24. This is an inference from Pound's remarks in ibid.

25. Record Book, p. 61.

26. Ibid., p. 63.

27. Pound, "Clements," p. 113.

28. Clements, unpublished reminiscence.

29. Pound, "Clements," p. 113; Clements, unpublished reminiscence.

30. Clements, unpublished reminiscence.

31. Ibid.

32. Record Book, pp. 63–69.

33. Pound, "Clements," p. 113.

34. Clements, unpublished reminiscence.

35. The thesis version, published in 1898, contains no mention of the quadrat method; Roscoe Pound and Frederic E. Clements, *The Phytogeography of Nebraska, I. General Survey* (Lincoln, Nebr.: Jacob North, 1898).

36. Frederic Clements's note typed onto the letter from W. C. Allee to Frederic Clements, December 8, 1933, Box 118, Clements Collection; Clements, unpublished reminiscence.

37. Record Book, p. 91.

38. An additional clue to this dating of the invention of the meter-plot is provided by Frederic Clements in his rough outline for a reminiscence on his friendship with Pound. His topic outline proceeds chronologically: "minors, degree; Commencement/Quadrat," which surely implies that the meter-plot was invented about the time of or shortly after Pound's doctoral commencement in June, 1897. Frederic Clements, "Pound Materials," an outline attached to Clements, unpublished reminiscence.

39. Charles E. Bessey, Report to the Regents of the University of Nebraska, December, 1898, Letterpress Books, 1898–1901, Bessey Papers.

40. In this discussion, I draw on a distinction between the theoretical and applied in science that is now widely used in the sociology of science. Theoretical problems derived from a basic disciplinary paradigm, without reference to social or economic concerns, while applied problems or technological problems derived from the application of knowledge to social and economic problems and were, from a disciplinary point of view, sui generis and ad hoc. The distinction does not imply that applied studies were more empirical than theoretical studies, or that results of

applied research could not be expressed in abstract, mathematical, and generalized form. This distinction is used, for example, by Joseph Ben-David, *The Scientist's Role in Society; A Comparative Study* (Englewood Cliffs, N.J.: Prentice-Hall, 1971), pp. 142–43.

41. I assume that book orders for these titles imply that the departmental and university libraries did not already have them. See C. E. Bessey to Wyer, Librarian, October 25, 1902, box 1902(S)–1903(L); C. E. Bessey to Lulau (?) & Company, October 9, 1901, Letterpress Books, 1902–1903; C. E. Bessey to Charles L. Smith, November 18, 1899, Letterpress Books, 1898–1901, Bessey Papers; Alexander von Humboldt, *De distributione geographica plantarum secundum coeli temperiem et altitudinem montium, prolegomena* (Lutetiae Parisiorum: in Libraria graeco-latino-germanica, 1817), ibid. and Aimé Bonpland, *Essai sur la géographie des plantes* (Paris: Librarie Levrault Schoell, 1807), Augustin de Candolle, *Regni vegetabilis systema naturale, sive Ordines, genera et sepecies plantarum secundum methodi naturalis normas digestarium et descriptarum*, 2 vols. (Parisiis: sumptibus sociorum Theuttel et Würtz, 1818–21), August Grisebach, *Bericht über die Leistungen in der Pflanzengeographie (und systematischen Botanik) während des Jahres 1843 (–53)* (Berlin: Nicolai, 1845–56), J. F. Schouw, *Grundzüge einer allgemein Pflanzengeographie . . . Aus dem Dänischen übersetzt* (Berlin: G. Reimer, 1823).

42. Another clue that the Bessey group could have been unacquainted with Drude's 1890 work before 1895 is that the Committee on Geographical Botany of the American Association for the Advancement of Science recommended Drude's 1890 *Handbuch* for the study of geographical botany only in the August, 1895, meeting; see the report of "Section G, A.A.A.S., Proceedings of the Section," *Botanical Gazette* 20 (September 1895):407–408. This implies a general recognition of Drude's *Handbuch* coming in 1895, and not earlier.

43. Close paraphrase of Roscoe Pound and Frederic Clements, "Vegetational Regions of the Prairie Province," *Botanical Gazette* 25 (1898):382–83.

44. Ibid., p. 383.

45. Roscoe Pound and Frederic Clements, "A Method of Determining the Abundance of Secondary Species," *Minnesota Botanical Studies* [1898] 2 (1898–1902):19.

46. Ibid.

47. Roscoe Pound and Frederic E. Clements, *The Phytogeography of Nebraska; I. General Survey*, second edition revised (Lincoln, Nebr.: Published by the Seminar, 1900), p. 60.

48. Ibid.

49. Pound and Clements, "A Method of Determining the Abundance," p. 22.

50. Clements, unpublished reminiscence.

51. Pound, "Clements," p. 113.

52. Clements, note on letter from Allee; Rodgers, *Coulter*, pp. 118–19.

53. Pound and Clements, *Phytogeography* (1900), p. 61.

54. Pound and Clements, "Method of Determining the Abundance," p. 20; Pound and Clements, *Phytogeography* (1900), p. 61.

55. Pound and Clements, *Phytogeography* (1900), p. 61.

56. In *Plant Succession*, however, Clements belatedly indicated that Charles Darwin had used a quadratlike method in his own research. There is no evidence that Pound and Clements were aware in 1897—1900 of Darwin's use of it. Clements, *Plant Succession*, p. 425.

57. Charles Darwin, *The Origin of Species by Means of Natural Selection*, facsimile edition of the first edition, edited by Ernst Mayr (New York: Atheneum, 1967), 4:83.

58. I am adopting Darwin's language in Darwin, *Origin of Species*, Mayr, ed., p. 84. My point of view on the meaning of the Darwinian revolution is indebted to Michael Ghiselin's reading of the *Origin*, in Michael T. Ghiselin, *The Economy of Nature and the Evolution of Sex* (Berkeley, Los Angeles, London: University of California Press, 1974), chap. 2, "The Legacy of the Stagirite, or, Teleology Old and New," pp. 16—48.

59. Frederic E. Clements, *Research Methods in Ecology* (Lincoln, Neb.: University Publishing, 1905), p. iii.

60. Clements, *Research Methods*, p. 20.

61. This is one of the many themes in Clarence J. Glacken, *Traces on the Rhodian Shore; Nature and Culture in Western Thought from Ancient Times to the End of the Eighteenth Century*, reprint edition (Berkeley, Los Angeles, London: University of California Press, 1976). The suitability of plant and animal organisms for their environment was related to the ancient Platonic doctrine called the Principle of Plenitude; see Arthur O. Lovejoy, *The Great Chain of Being; A Study in the History of Ideas*, reprint edition (Cambridge, Mass.: Harvard University Press, 1964). The chain of being was a scale of continuous qualities, or categories, rather than a scale of mathematical degrees of the same quality; see Lovejoy, *The Great Chain of Being*.

62. Clements, *Research Methods*, p. 20.

63. Ibid., p. 37.

64. Ibid., p. 97.

65. Ibid., p. 101.

66. Ibid.

67. Ibid., p. 163.

68. Ibid.

69. Ibid.

70. Ibid., p. 164.

71. Ibid., pp. 164—65.

72. Ibid., pp. 173—75.

4. Frederic Clements's Theory of Plant Succession

1. Edith S. Clements, *Adventures in Ecology; Half a Million Miles . . . From Mud to Macadam* (New York: Pageant Press, 1960), p. 226. Edith Clements's charming memoir presents difficulties of interpretation. The style is virtually novelistic, complete with character sketches and quoted dialogue. Frederic refers, apparently, to the writing of the book in 1934; Frederic E. Clements to W. P. Taylor,

October 12, 1934, box 119, Dr. Edith S. Clements and Frederic E. Clements Collection, Division of Rare Books and Special Collections, The Library, University of Wyoming, Laramie, Wyoming; hereafter, Clements Collection. Based on analysis of the style, I infer that the episodic chapters were developed by Mrs. Clements from travel diaries and that the dialogue, therefore, is close to original conversation.

2. Raymond J. Pool, "Frederic Edward Clements," *Ecology* 35 (1954):109.

3. For biographical memoirs, see Joseph Ewan, "Clements, Frederic Edward," *Dictionary of Scientific Biography*, ed. Charles Coulton Gillispie (New York: Charles Scribner's Sons, 1973), pp. 317–18; Pool, "Clements"; Roscoe Pound, "Frederic E. Clements As I Knew Him," *Ecology* 35 (1954):112–13; Paul B. Sears, "Clements, Frederic Edward," *Dictionary of American Biography*, ed. Edward T. James (New York: Charles Scribner's Sons, 1973), Supplement Three, 1941–45, pp. 168–70.

4. Frederic E. Clements, unpublished memoir, enclosed in Frederic E. Clements to Sayre, January 17, 1945, box 125, Clements Collection.

5. On their joint reading, see Pound, "Frederic E. Clements As I Knew Him"; on Pound's activities and his law career in the Nebraska years, see David Wigdor, *Roscoe Pound, Philospher of Law* (Westport, Conn. and London: Greenwood Press, 1974), pp. 69–132.

6. Pool, "Clements," p. 109.

7. Edith S. Clements, *Adventures in Ecology*, pp. 14–15.

8. Wigdor, *Roscoe Pound*, pp. 129–33.

9. Frederic E. Clements, *Plant Succession: An Analysis of the Development of Vegetation* (Washington, D.C.: Carnegie Institution, 1916), p. iii.

10. Edith S. Clements, *Adventures in Ecology*, p. 65.

11. See chap. 5, below.

12. A. G. Tansley, "Development of Vegetation," *Journal of Ecology* 4 (1916):198–204; see also the characterization of Tansley's interest in Clements's theory in Harry Godwin, "Sir Arthur Tansley: The Man and His Subject," *Journal of Ecology* 65 (1977):1–26.

13. Frederic E. Clements, *The Development and Structure of Vegetation*, University of Nebraska, Botanical Survey of Nebraska, 7, Studies in the Vegetation of the State, 3 (Lincoln, Nebr.: Published by the Seminar, 1904); *Research Methods in Ecology* (Lincoln: University Publishing, 1905); *Plant Succession: An Analysis of the Development of Vegetation* (Washington, D.C.: Carnegie Institution, 1916). The major themes of Clementsian ecology, discussed in the previous paragraph in the text, are not expressed precisely the same in the three titles and, of course, there is development of themes from 1904 to 1916. Thus, in *Development and Structure of Vegetation*, Clements believed that the formation was an entity, but he wrote of it as comparatively similar, not ontologically equivalent, to the individual organism: phenomena peculiar to the formation would "be clearer if we consider vegetation as an entity" (p. 5). While the organismic quality of the formation may have been ambiguous here, Clements nevertheless believed that the formation was more than a mechanical association of plants. Similarly, regarding the progressive

nature of succession, Clements wrote in *Development and Structure* of changes in the formation as "rhythmic and progressive," but did not define progressive as inevitable (p. 6). Rather, he allowed for "imperfect" succession, when stages in the series were omitted or a normally earlier stage appeared later (p. 122). Vegetation developed toward stabilization "as a general principle," rather than as an iron necessity (p. 122). Regardless of the ambiguities in 1904, the following year, Clements had made up his mind and we find the central themes of his work stated dogmatically.

Critical reviews—not to say histories—of the concept of formation and succession are provided by Robert H. Whittaker, "Classification of Natural Communities," *The Botanical Gazette* 28 (1962):1–239, and Robert P. McIntosh, "The Continuum Concept of Vegetation," *The Botanical Review* 33 (1967):130–87; David W. Shimwell, *The Description and Classification of Vegetation* (Seattle: University of Washington Press, 1971), pp. 122–26. On the concept of succession, see the papers reprinted in Frank B. Golley, ed., *Ecological Succession*, Benchmark Papers in Ecology 15 (Stroudsburg, Pa.: Dowden, Hutchinson, & Ross, 1977), esp. W. H. Frury and I. C. T. Nisbet, "Succession," pp. 287–324.

14. Clements, *Plant Succession*, p. 4. On the recapitulation analogy, see Stephen Jay Gould, *Ontogeny and Phylogeny* (Cambridge, Mass.: Belknap Press of Harvard University Press, 1977).

15. Clements, *Plant Succession*, p. 3.

16. Ibid.

17. The paraphrase is from ibid., p. 6; on the origin of the mechanical equilibrium metaphors in Pound and Clements, see Ronald Tobey, "Theoretical Science and Technology in American Ecology," *Technology and Culture* 17 (October 1976): 718–28. A general account of equilibrium concepts, which does not, however, discuss ecology, is provided by Cynthia E. Russett, *The Concept of Equilibrium in American Social Thought* (New Haven: Yale University Press, 1966). I here present the concept of organism and equilibrium as contradictory, which, I believe, they were in Clements's work. Russett has shown that, despite a duality in concepts, in twentieth-century animal physiology the concept of the animal organism, conceived as a functional whole, depended upon mechanical metaphors borrowed from physics and chemistry; Russett, *Concept of Equilibrium*, pp. 4, 19, 21. For a conceptual clarification of early twentieth-century philosophical organicism, see D. C. Phillips, "Organicism in the Late Nineteenth and Early Twentieth Centuries," *Journal of the History of Ideas* 31 (July–September 1970):413–32, which does not, however, touch on Clements's organicism. Phillips sees a conflict between organicism and equilibrium theory.

See also Frank N. Egerton, "Changing Concepts of the Balance of Nature," *The Quarterly Review of Biology* 48 (1973):322–50.

18. Clements, *Plant Succession*, p. 6.

19. See chap. 7 below.

20. Clements, *Plant Succession*, p. 6.

21. Ibid.

22. Ibid., p. 3.

23. Tansley, "Development of Vegetation," p. 198.

24. Clements, *Plant Succession*, p. 145.

25. Ibid.

26. For an additional summary of Clements's system, see R. H. Whittaker, "Recent Evolution of Ecological Concepts in Relation to the Eastern Forests of North America," reprinted in Frank N. Egerton, ed., *History of American Ecology* (New York: Arno Press, 1977), pp. 340–43.

27. Clements, *Research Methods in Ecology*, p. 5.

28. Edith S. Clements, *Adventures in Ecology*, pp. 14–16.

29. American Association for the Advancement of Science, *Proceedings* 50 (1901):332–33.

30. John Phillips, "Succession, Development, the Climax, and the Complex Organism: An Analysis of Concepts; Part III. The Complex Organism," *Journal of Ecology* 23 (1935):493. On Phillips's correspondence with Clements, see the account in John Phillips, "A Tribute to Frederic E. Clements and His Concepts in Ecology," *Ecology* 35 (1954):115. Donald Worster, *Nature's Economy: The Roots of Ecology* (San Francisco: Sierra Club Books, 1977), p. 212, mentions Spencer but not Ward as the source of Clements's organicism.

31. Lester F. Ward, *Dynamic Sociology, or Applied Social Science*, 2 vols., second edition, revised (New York: D. Appleton, 1897); Herbert Spencer, *The Principles of Biology*, 2 vols. (Osnabruck: Otto Zeller, reprint of 1898 edition, 1966); Herbert Spencer, *Principles of Sociology*, vol. 1 (Osnabruck: Otto Zeller, reprint of 1904 edition, 1966); Pound, "Frederic E. Clements As I Knew Him," p. 113.

32. Wigdor, *Roscoe Pound*, p. 111; E. A. Ross to Frederic E. Clements, January 13, 1936, box 121, Clements Collection; Roscoe Pound, "A New School of Jurists," pp. 249–66, *University Studies* (University of Nebraska) 4 (July 1904), esp. pp. 265–66.

33. Herbert Spencer, *The Study of Sociology*, first edition, 1873 (New York: D. Appleton, reprint edition, 1902), p. 301.

34. Ward, *Dynamic Sociology*, 1:35.

35. Edward J. Pfeifer, "The Genesis of American Neo-Lamarckism," *Isis* 56 (1965):156–76; Hamilton Cravens, *The Triumph of Evolution: American Scientists and the Heredity-Environment Controversy, 1900–1941* (Philadelphia: University of Pennsylvania Press, 1978), pp. 35–39.

36. John S. Haller, Jr., *Outcasts from Evolution: Scientific Attitudes of Racial Inferiority, 1859–1900*, reprint edition (New York: McGraw-Hill, 1975), pp. 98–99, 187–202.

37. Clements, *Research Methods*, p. 148.

38. Frederic E. Clements to Merriam, June 24, 1925, box 113, Clements Collection; Frederic E. Clements to Frederic W. (Taylor?), April 30, 1942, box 124, Clements Collection.

39. Frederic E. Clements to Oscar Drude, first page of letter missing, date certainly after 1919, box 111 Clements Collection.

40. The theme of idealism in German plant geography has been briefly treated by Michael T. Ghiselin, *The Economy of Nature and the Evolution of Sex* (Berkeley, Los Angeles, London: University of California Press, 1974), pp. 28–32. Worster, *Nature's Economy*, pp. 198–202, 206–9, completely misses the importance of Drude's plant geography, and his account of the origin of American plant ecology,

which is not derived from manuscript sources, should not be relied upon.

41. The basic history of nineteenth-century idealism is well known and presented in many textbooks, so no bibliography of it is needed here. On my text, see Herbert W. Schneider, *A History of American Philosophy*, reprint edition (New York and London: Columbia University Press, 1963), pp. 375—415; Robert N. Manley, *Centennial History of the University of Nebraska, I. Frontier University (1869—1919)* (Lincoln: University of Nebraska Press, 1969), p. 131; Edgar Lenderson Hinman, *The Physics of Idealism* (Lincoln: State Journal Company, 1906). I have been unable to discover whether Frederic Clements and his colleague, Hinman, were active friends; we must assume that in the same university, however, they knew each other.

42. Humboldt's letter is quoted from Alexander von Humboldt to Caroline von Wolzogen, 1806, in Karl C. Bruhns, *Life of Alexander von Humboldt*, 2 vols., trans. Jane and Caroline Lassell (London: Longmans, Green, 1873), 1:359. On Humboldt's youthful philosophical proclivities, see Bruhns, 1:185, 195—97, 200—8, and Hanno Beck, *Alexander von Humboldt*, 2 vols. (Wiesbaden: Franz Steiner Verlag GMBH, 1959—61), 1:46, 60—61, 66—67. On Humboldt's involvement with French scientists, see Robert Ave-Lallemant, "Alexander von Humboldt; Sojourn in Paris from 1808 to 1826," in Bruhns, 2:1—71; Beck, 2:12—76; Maurice Crosland, *The Society of Arcueil: A View of French Science at the Time of Napoleon I* (Cambridge, Mass.: Harvard University Press, 1967), pp. 104—13; L. Kellner, *Alexander von Humboldt* (London: Oxford University Press, 1963), pp. 78—88. On Humboldt's relationship with Goethe, see Bruhns, 1:161—178, and Beck, 1:65—68. On Humboldt's inspiration from Karl Willdenow and George Förster, see Beck, 2:66, and Bruhns, 1:84—85.

On Goethe's romantic science, see his *The Metamorphosis of Plants*, reprinted in Johann w. Von Goethe, *Goethe's Botany; The Metamorphosis of Plants (1790) and Tobler's Ode to Nature (1782)*, trans. Agnes Arber (Waltham, Mass.: Chronica Botanica [vol. 10, no. 2], 1946). Goethe's doctrines of original type and organistic holism are discussed by Timothy Lenoir, "Generational Factors in the Origin of *Romantische naturphilosophie*," *Journal of the History of Biology* 11 (1978): 64—65, Rudolf Magnus, *Goethe as a Scientist*, trans. Heinz Norden (New York: Henry Schuman, 1949), pp. 37—80, 94—100, and H. B. Nisbet, *Goethe and the Scientific Tradition* (London: Institute of German Studies, 1972), pp. 6—22.

I am indebted to Frank Egerton for criticism and encouragement in my efforts to evaluate Humboldt's position in plant geography.

43. Immanuel Kant, *Critique of Judgment*, sec. 66, "The Principle on which the intrinsic finality [relation to ends] in organisms is estimated," *The Philosophy of Kant; Immanuel Kant's Moral and Political Writings*, ed. Carl J. Friedrich, trans. James C. Meredity (New York: Modern Library, 1949), p. 318.

44. Kant, *Critique of Judgment*, p. 321.

45. Beck, *Humboldt*, 1:60.

46. Alexander von Humboldt and Aimé Bonpland, *Essai sur la géographie des plantes* (Paris: Libraire Levrault Schoell, 1807), preface, vi; Ghiselin, *Economy of Nature*, p. 29; Worster, *Nature's Economy*, pp. 133—36, 183.

47. Alexander von Humboldt, *Aspects of Nature*, trans. (from the Ger-

man) Mrs. Sabine, 3d ed. first English edition, 2 vols. (London: Longman, Brown, Green, and Longmans, 1849), 1:13.

48. Alexander von Humboldt, *Cosmos, A Sketch of the Universe*, trans. by E. C. otto, 2 vols. (New York: Harper, 1850), 1:17.

49. Ibid., p. 21. On Goethe's views, see the references in n. 42, above.

50. Humboldt, *Aspects of Nature*, 1:9—10.

51. August Grisebach's proposal of the term "formation" came in "Über den Einfluss das Klima auf die Begrenzung der natürlichen Floren," reprinted in *Gesammelte Abhandlungen und kleinere Schriften zur Pflanzengeographie* (Leipzig: Verlag von Wilhelm Engelmann, 1880), pp. 1—29; August Grisebach "Pflanzengeographie und Botanik," in Karl C. Bruhns, ed., *Alexander von Humboldt*, 3 vols. (Leipzig: Brockhaus, 1872), 3:248.

52. See Clements, *Plant Succession*, pp. 116—17.

53. Frank Dawson Adams, *The Birth and Development of the Geological Sciences*, first edition, 1938 (New York: Dover, 1954), pp. 216—23, 271—73; Roy Porter, *The Making of Geology: Earth Science in Britain, 1660—1815* (Cambridge [England]: Cambridge University Press, 1977), pp. 167—68; Cecil Schneer, "The Rise of Historical Geology in the Seventeenth Century," *Isis* 45 (1954): 256—68; William Smith, *Stratigraphic System of Organized Fossils*, with reference to the specimens of the original geological collection in the British Museum: explaining their state of preservation and their use in identifying the British strata (London: E. Williams, 1817).

54. Leonard G. Wilson, *Charles Lyell; The Years to 1841: The Revolution in Geology* (New Haven and London: Yale University Press, 1972), pp. 248—55, 259—60; Charles Lyell, *Principles of Geology*, 2 vols. (London: J. Murray, 1830—32); Charles Lyell, *Elements of Geology* (London: Murray, 1838), pp. 284—85.

55. Lyell discussed this question briefly with regard to the attempt of the French geologist, Alexander Brongniant, to estimate numbers of species in the carboniferous era; see Lyell, *Principles of Geology*, 1:101. Also see Lyell's use of the Humboldtian framework in his analysis of shells in *Elements of Geology*, pp. 289—290.

56. Clements, *Plant Succession*, pp. 344—422.

57. For a brief biography of Drude, see Rudolph Zaunick, "Drude, Carl Georg Oscar," *Neue deutschen Biographie* 4 (Berlin: Duncker & Humblot, 1957), p. 138. For Drude's own statement of his relationship to Grisebach, see the foreword in Oscar Drude, *Handbuch der Pflanzengeographie* (Stuttgart: J. Engelhorn, 1890), esp. p. ix. The *Handbuch* was dedicated to Grisebach.

58. Drude, *Handbuch*, pp. 223—26.

59. Ibid., p. 225.

60. Ibid., pp. 287—304, 441—42.

61. Ibid., pp. 108—9.

62. Ibid., pp. 100—103. This language of plant warfare antedated Darwin's *Origin of Species*, being most widely available to botanists in Alphonse de Candolle, *Géographie botanique raisonnée*, 2 vols. (Paris: V. Masson, 1855).

63. Drude listed the major classes of formations in the *Handbuch*, p. 229, and discussed them, pp. 230—326.

64. For example, see de Candolle's own comments, *Géographie bota-*

nique, 1, preface, p. vi, and Drude's remarks, *Handbuch*, "Einleitung," p. 7.

65. De Candolle, *Géographie botanique*, 1:v.

66. Ibid., p. 2.

67. De Candolle made the problem of the division of vegetation into "natural regions," the subject of an entire chapter, "XXV. De la division de surfaces terrestres en régions naturelles," ibid., 2:1298–1310. For the critique of Humboldt's theory, see ibid., pp. 1299–1300, 1304–6.

68. Ibid., p. 1309.

69. Frank N. Egerton, "Humboldt, Darwin, and Population," *Journal of the History of Biology* 3 (Fall, 1970):325–60.

70. Charles Darwin, *The Origin of Species By Means of Natural Selection*, 6th ed. (London: John Murray, 1900), chap. 12, p. 498.

71. Ibid., chap. 10, p. 412.

72. Biographical details are provided by D. Müller, "Warming, Johannes Eugenius Bülow," *Dictionary of Scientific Biography*, ed. Charles Coulston Gillispie (New York: Charles Scribner's Sons, 1976), 14:181–82. Eugene Warming, *Lehrbuch der ökologischen Pflanzengeographie; eine Einführung in die Kenntnis der Pflanzenvereine*, trans. Emil Knoblaugh (Berlin: Gebrîsder Borntraeger, 1896).

73. Warming, *Lehrbuch*, p. 10.

74. Ibid., pp. 105–10.

75. Ibid., p. 110.

76. Eug. Warming, *Oecology of Plants*, trans. P. Groom and I. B. Balfour (Oxford: Clarendon Press, 1909), p. 95.

77. Warming, *Lehrbuch*, p. 352.

78. Warming, *Oecology*, p. 349.

79. Warming, *Lehrbuch*, p. 364.

80. Warming, *Oecology*, p. 359.

81. Warming, *Lehrbuch*, p. 372; Warming, *Oecology*, p. 365.

82. Warming, *Oecology*, p. 348.

83. Ibid.

84. See Warming's references, *Oecology*, p. 350; H. C. Cowles, "The Ecological Relations of the Vegetation on the Sand Dunes of Lake Michigan," *Botanical Gazette* 27 (1899):95–117, 167–202, 281–308, 361–91.

85. The entire discussion is provided in Warming, *Lehrbuch*, pp. 352–63, and *Oecology*, pp. 349–59.

86. See the testimony in Cowles, "Sand Dunes," p. 97.

87. Ibid.

88. "News," *Botanical Gazette* 26 (1898):223.

89. Cowles, "Sand Dunes," pp. 97–98.

90. See Cowles, "Sand Dunes;" H. C. Cowles, "The Physiographic Ecology of Chicago and Vicinity; A study of the Origin, Development, and Classification of Plant Societies," *Botanical Gazette* 31 (1901):73–108, 145–82; H. C. Cowles, "The Influence of Underlying Rocks on the Character of the Vegetation," *Bulletin*, American Bureau of Geography, 2 (1901):163–76, 376–88; and H. C. Cowles, "The Causes of Vegetative Cycles," *Botanical Gazette* 51 (1911):161–83.

91. Cowles, "Sand Dunes," p. 111.

92. Ibid.

93. Ibid.; the reference to causal agencies is my interpretation of the passage in Cowles.

94. Cowles, "Physiographic Ecology of Chicago," p. 81.

95. Cowles, "Causes of Vegetative Cycles," pp. 168–73.

96. Ibid., p. 172.

97. H. C. Cowles, review of Roscoe Pound and Frederic E. Clements, *The Phytogeography of Nebraska*, in *Botanical Gazette* 25 (1898):372. The picture I am drawing of these two schools of ecology is sketched, without its intellectual consequences being filled in, by Paul B. Sears, *The Living Landscape* (New York: Basic Books, 1966), pp. 76–80. Sears, a doctoral student of Cowles, also taught briefly at Nebraska.

5. The Life Cycle of Grassland Ecology

1. The research tests underlying some of the analyses in this chapter were discussed in my article, "American Grassland Ecology, 1895–1955: the Life Cycle of a Professional Research Community," in Frank N. Egerton, ed., *History of American Ecology* (New York: Arno Press, 1977), p. 45, plus eleven pages of figures. For a discussion of the first step in Crane's scenario, see Diana Crane, *Invisible Colleges; Diffusion of Knowledge in Scientific Communities* (Chicago and London: University of Chicago Press, 1972), pp. 35 ff.

2. Ibid., pp. 51 ff. On Braun's early adoption of Clementsianism, see the correspondence between her and Frederic Clements in the Dr. Edith S. Clements and Dr. Frederic Clements Collection, Division of Rare Books and Special Collections, The Library, University of Wyoming, Laramie, Wyoming, boxes 118, 120. Hereafter, Clements Collection.

3. See Gerard Lemaine, et al., eds., *Perspectives on the Emergence of Scientific Disciplines* (Paris and The Hague: Mouton; Chicago: Aldine, 1976), for a recent collection of papers, pp. 3–6 for a discussion of Mendelism; David O. Edge and Michael J. Mulkay, *Astronomy Transformed; The Emergence of Radio Astronomy in Britain* (New York: John Wiley, 1976), especially the discussion in chap. 10, "Some Sociological Implications," pp. 368–69.

4. Thomas Kuhn, *The Copernican Revolution, Planetary Astronomy in the Development of Western Thought*, reprint edition (New York: Vintage Books, 1959); Gerald Holton, "The Roots of Complementarity," pp. 115–61, in *Thematic Origins of Scientific Thought, Kepler to Einstein*, reprint edition (Cambridge, Mass.: Harvard University Press, 1975); Lewis Feuer, "Niels Bohr: The *Ekliptika* Circle and the Kierkagaardian Spirit," pp. 199–257, in *Einstein and the Conflict of Generations* (New York: Basic Books, 1974); Edge and Mulkay, *Astronomy Transformed*, pp. 359–64, and fig. 10.1, p. 382.

5. On the normal science stage and puzzle-solving, see Kuhn, *Structure of Scientific Revolutions*, pp. 23–42; and Crane, *Invisible Colleges*, pp. 26–31. Larry Laudan presents a non-Kuhnian theory of scientific progress, in which problem-

solving, rather than puzzle-solving, is the main feature of research; see Larry Laudan, *Progress and Its Problems: Toward a Theory of Scientific Growth* (Berkeley, Los Angeles, London: University of California Press, 1977).

6. Crane, *Invisible Colleges*, p. 172.

7. Kuhn, *Structure of Scientific Revolution*, pp. 66–76; Crane, *Invisible Colleges*, p. 172. It is a weakness of Crane's work, that her alternative scenario of "exhaustion" is not discussed at length, but only raised as a possibility in fig. 1 (p. 172) of her book.

8. See Warren Hagstrom's now classic work on the sociology of science, *The Scientific Community* (New York: Basic Books, 1965). For a review of empirical findings of sociologists of science, see the discussion by Edge and Mulkay, *Astronomy Transformed*, chap. 10. For a discussion of the "normative" and the "interpretive" viewpoints in the sociology of science, see John Law, "Theories and Methods in the Sociology of Science: An Interpretive Approach," pp. 220–31, in Lemaine, et al., eds., *Perspectives on the Emergence of Scientific Disciplines*.

9. Richard J. Storr, *Harper's University, The Beginnings: A History of the University of Chicago* (Chicago and London: University of Chicago Press, 1966), p. 68.

10. William Michael Murphy and D. J. R. Bruckner, eds., *The Idea of the University of Chicago; Selections from the Papers of the First Eight Chief Executives of the University of Chicago, 1891–1975* (Chicago and London: University of Chicago Press, 1976), p. 16.

11. Ibid., p. 134.

12. Andrew Denny Rodgers, III, *John Merle Coulter: Missionary in Science* (Princeton: Princeton University Press, 1944), pp. 38–40, 124–25, 236–37, 292–97.

13. John Merle Coulter and Merle C. Coulter, *Where Evolution and Religion Meet* (New York: Macmillan, 1924), p. 11.

14. See Ronald C. Tobey, *The American Ideology of National Science, 1919–1930* (Pittsburgh: University of Pittsburgh Press, 1971), esp. chaps. 5 and 6.

15. Murphy and Bruckner, eds., *The Idea of the University of Chicago*, p. 17.

16. On proposals for study of the Nebraskan sand hills after Bessey's passing, see Raymond J. Pool, chairman, Department of Botany, to Samuel Avery, chancellor, University of Nebraska, November 4, 1924, Pool File, Correspondence with academic departments, Office of the Chancellor (Papers), The University Archives, The University of Nebraska, Lincoln.

17. "News and Notes," *Botanical Gazette 21* (March 1896):183.

18. "News," ibid., 28 (July 1899):78–79.

19. Ibid., 26 (September 1898):223.

20. Ibid.; ibid., 30 (August 1900):142; ibid., 32 (July 1901):74–75; ibid., 33 (May 1902):399–400; ibid., 34 (October 1902):320; ibid., 36 (September 1903):238; ibid., 37 (March 1904):239–40; ibid., 42 (September 1906):239; ibid., 44 (July 1907):79.

21. Clements first visited the Pike's Peak region for ecological study in August 1900, and took students there with him the following years; ibid. 30 (August

1900):144, and ibid. 33 (July 1901):74–75. See also the charming account, Edith S. Clements, *Adventures in Ecology: Half a Million Miles . . . From Mud to Macadam* (New York: Pageant Press, 1960), pp. 14–17.

22. "News," *Botanical Gazette* 33 (May 1902):400.

23. Alfred Dachnowski, "The International Phytogeographical Excursion of 1913 and Its Significance to Ecology in America," *Journal of Ecology* 2 (1914):240. In fact, Dachnowski refers to "the commanding influence that has justly been exercised by the Chicago school of ecology."

24. See Ronald Tobey, "Theoretical Science and Technology in American Ecology," *Technology and Culture* 17 (October 1976):718–28.

25. Kuhn, in *Structure of Scientific Revolutions*, uses "scientific community" to imply a "collaborative community."

26. Crane, *Invisible Colleges*, p. 49.

27. David F. Costello, *The Prairie World* (New York: Thomas Y. Crowell, 1969), p. v.

28. H. Gilman McCann, *Chemistry Transformed: The Paradigmatic Shift from Phlogiston to Oxygen* (Norwood, N. J.: Ablex, 1978), pp. 11, 108–9, 111.

29. Kuhn's remark that allegiance to a paradigm steers a scientist's choice of research problems to ones that can be solved in terms of the paradigm implies that the scientist will not even be concerned with the problems that critics hold against the paradigm. See Kuhn, *Structure of Scientific Revolutions*, p. 37.

30. Derek de Solla Price, "A General Theory of Bibliometric and Other Cumulative Advantage Processes," *Journal of the American Society for Informational Science* 27 (1976):292–306.

31. See chap. 7.

32. See Kuhn's discussion of textbooks, *Structure of Scientific Revolutions*, pp. 11–12, 43, 137–43, 165.

6. A. G. Tansley

1. The quotation, and the characterization of Tansley are derived from Sir Harry Godwin, "Sir Arthur Tansley: The Man and the Subject," *Journal of Ecology* 65 (1977): 1–26. Godwin testifies that he (Godwin) was also inspired by Clements's writings, spending most of the year 1921 reading them; ibid., p. 14.

2. I have relied upon the study by P. Lowe for my understanding of the sociological situation in English natural history at the end of the nineteenth century; see P. D. Lowe, "Amateurs and Professionals: The Institutional Emergence of British Plant Ecology," *Journal of the Society for Bibliography of Natural History* 7 (1976): 517–535. I appreciate the favor of Mr. Lowe of introducing me to his paper. See also, David Elliston Allen, *The Naturalist in Britain; A Social History* (London: Allen Lane, 1976).

3. Godwin suggests that Oliver's work enticed Tansley into ecology; Godwin, "Sir Arthur Tansley," p. 4.

4. For an account of the 1913 excursion, see Alfred Dachnowski, "The

International Phytogeographical Excursion of 1913 and Its Significance to Ecology in America," *Journal of Ecology 2* (1914):237—45. The letters of 1911 of Dr. Edith Clements, Frederic Clements's wife, are filled with personal and incidental information about the botanists the Clementses met in Britain, including the Tansleys, the Henry Cowles, and the Brittons (of Columbia University); Edith G. Clements Papers, Nebraska State Historical Society, State Archives Division, Manuscript Division, Lincoln, Nebraska.

5. Robert L. Burgess, "The Ecological Society of America; Historical Data and Some Preliminary Analyses," in Frank N. Egerton, ed., *History of American Ecology* (New York: Arno Press, 1977), 24 pp.

6. Quoted in Godwin, "Sir Arthur Tansley," p. 7.

7. F. F. Blackman and A. G. Tansley, "Ecology in Its Physiological and Phytogeographical Aspects," *The New Phytologist 4* (November 1905):199.

8. W. C. Smith, "[Report on] the Central Committee for the Survey and Study of British Vegetation," ibid. (December 1905):255.

9. A. G. Tansley, "The Problems of Ecology," *The New Phytologist 3* (1904):192.

10. Godwin, "Sir Arthur Tansley," p. 14.

11. A. G. Tansley, *Practical Plant Ecology*, 2d ed. (London: George Allen & Unwin, 1926), pp. 18—19.

12. Ibid., p. 19.

13. Blackman and Tansley, "Ecology," p. 247.

14. Ibid., p. 245.

15. Tansley, *Practical Plant Ecology*, p. 98.

16. Ibid., p. 127.

17. Arthur G. Tansley, *The New Psychology* (London: George Allen & Unwin, 1922), p. 33.

18. See Stephen Jay Gould, *Ontogeny and Phylogeny* (Cambridge, Mass.: Belknap Press of Harvard University Press, 1977). Darwin made the dictum one of three classes of evidence for the descent of man, in *The Descent of Man and Selection in Relation to Sex* (New York: The Modern Library Reprint of 1871 Edition, n.d.), chap. 1, pp. 398—400, where Darwin drew on the verbal support of Thomas H. Huxley.

19. Tansley, *Practical Plant Ecology*, pp. 21—25.

20. Ibid., p. 23.

21. For a comparison of openly idealistic language in the interpretation of biological phenomena, see C. Lloyd Morgan, *The Interpretation of Nature* (Bristol [England]: J. W. Arrowsmith, 1905). The same language was employed by the well-known philosophical organicists of the 1920s, Samuel Alexander and Alfred North Whitehead. For a representative anthology, with introductory essays and brief bibliographies, see Douglas Browning, ed., *Philosophers of Process* (New York: Random House, 1965). See also the recent utilization of Whitehead's and Bergson's views, Mîlîc Câpek, *The Philosophical Impact of Contemporary Physics* (Princeton: Van Nostrand, 1961).

22. For example, see Tansley, *Practical Plant Ecology*, p. 127.

23. Arthur G. Tansley, "The Classification of Vegetation and the Concept of Development," *Journal of Ecology* 8 (1920):124.

24. Ibid., pp. 43—46, 48—50.

25. Ibid., p. 163.

26. Ibid., pp. 37—38. The comparison of ecological strata in the forest and class strata in human society was ideologically pointed, as we shall see in the final section of this chapter. Even Godwin remarks about Tansley's opposition to social leveling; Godwin, "Sir Arthur Tansley," p. 25.

27. Ibid. p. 23.

28. Ibid., p. 24.

29. Arthur G. Tansley, "The Development of Vegetation," *Journal of Ecology* 4 (1916):198.

30. Ibid., p. 203.

31. Frederic E. Clements, *Plant Succession; An Analysis of the Development of Vegetation* (Washington, D.C.: Carnegie Institution, 1916), p. 3.

32. Ibid., preface, p. iii; italics added.

33. Quoted in A. G. Tansley, "Classification of Vegetation, p. 118.

34. Tansley, "Development of Vegetation," p. 203.

35. Ibid. Obviously, this chapter on Tansley's views is not intended to be a general narrative of the history of British plant ecology. Nevertheless, we can point out here that Tansley's opposition to Clements's monoclimax theory was developed together with his colleague, C. E. Moss, whose work subsequently influenced Clements's *Plant Succession*. See David W. Shimwell, *The Description and Classification of Vegetation* (Seattle: University of Washington Press, 1971), pp. 44—45, 47—49, for a summary of the English tradition. On the historical geography of Great Britain, see William G. Hoskins, *The Making of the English Landscape* (London: Hodder and Stoughton, 1955).

36. Tansley, "Classification of Vegetation, and the Concept of Development,' *Journal of Ecology* 8 (1920):124. and p. 121.

37. Ibid.

38. A. G. Tansley, "The Use and Abuse of Vegetational Concepts and Terms," *Ecology* 16 (1935):284—307; F. E. Clements to A. G. Tansley, April 28, 1935, box 125, Dr. Edith G. Clements and Frederic E. Clements Collection, Division of Rare Books and Special Collections, University Library, University of Wyoming, Laramie, Wyoming; hereafter, Clements Collection. My italicization added to the title in the text.

39. John Phillips, "Succession, Development, the Climax, and the Complex Organism: An Analysis of the Concepts," *Journal of Ecology* 22 (1934):554—71; 23 (1935):210—46, 488—508.

40. See the *Report of the Proceedings*, the 5th International Botanical Congress, 1930, F. T. Brooks and T. F. Chipp, eds. (Cambridge: University Press, 1931); on the nomenclature commission, which Tansley originally opposed, see A. G. Tansley to F. Clements, September 25, 1930; Tansley to Clements, November 10, 1930; Tansley to Clements, March 19, 1931, box 116, Clements Collection.

41. A hint of the pressures on Tansley and the difficulty in completing

Vegetation of the British Isles appeared in A. G. Tansley to F. Clements, April 17, 1934, box 125, Clements Collection.

42. Blackman and Tansley, "Ecology," p. 247.

43. Ibid., p. 251.

44. Ibid., pp. 247—52.

45. Tansley, "Classification of Vegetation," p. 126; the quotation is the key proposition in H. A. Gleason's paper, "The Structure and Development of the Plant Association," *Bulletin of the Torrey Botanical Club* 44 (October 1917): 463—81.

46. Tansley, "Classification of Vegetation," p. 125.

47. Gleason, "Structure and Development," pp. 464, 473, 484.

48. Ibid., p. 473.

49. Henry A. Gleason, "The Individualistic Concept of the Plant Association," *Bulletin of the Torrey Botanical Club* 53 (January 1926):26.

50. Arthur G. Tansley, "Succession: The Concept and Its Value," in B. M. Duggar, ed., *Proceedings of the International Congress of Plant Sciences*, 2 vols. (Menasha, Wisc.: Collegiate Press, George Banta, 1929)1:678.

51. Ibid., p. 679.

52. Ibid., pp. 686—87; see also, p. 679.

53. See n. 39.

54. Tansley, "Use and Abuse," p. 284.

55. Clements, *Plant Succession*, preface, p. iii.

56. Tansley, "Use and Abuse," passim.

57. C. Lloyd Morgan, *Emergent Evolution* (London: Williams and Norgate, 1923).

58. Ibid., p. 38.

59. Ibid., p. 39.

60. Clements, *Plant Succession*, p. 100.

61. Ibid., p. 3.

62. Ibid.

63. Ibid., p. 5.

64. Ibid., p. 7.

65. Ibid., p. 100.

66. Ibid.

67. Gleason, "Structure and Development," pp. 469—70.

68. See Clements's discussion of invasion, *Plant Succession*, pp. 75—76; see also, F. E. Clements to Billy (Allred?), January 17, 1926, box 114, Clements Collection, where Clements refuses to accept the notion that *all* transitions are continuous and that one meets in nature discrete boundaries of at least *some* complex organisms. For a review of the issue, see Robert P. McIntosh, "The Continuum Concept of Vegetation," *The Botanical Review* 33 (1967):130—87.

69. Tansley, "Use and Abuse," p. 300. Jack Major, "Historical Development of the Ecosystem Concept," pp. 9—22, in George M. Van Dyne, *The Ecosystem Concept in Natural Resource Management* (New York: Academic Press, 1969).

70. Shortly after the publication of Tansley's "Use and Abuse" article,

an English translation was published of A. I. Oparin's essay, *The Origin of Life*, which presented a speculative biochemical theory of the origination of primitive organic molecules and their evolution. It is not clear that Tansley was familiar with the earlier Russian version of the essay, but certain aspects of the theory are comparable to his system approach. See A. I. Oparin, *The Origin of Life*, trans. (from the Russian) Sergiuj Morgulis, first edition, 1938 (New York: reprint edition, Dover, 1953).

71. H. Levy, *The Universe of Science*, 2d ed. (London: Watts, 1938), p. 49.

72. Ibid., pp. 45–67, passim.

73. Ibid., p. 70.

74. Ibid., pp. 67–74. Levy did not deal directly with the cosmological consequences of Einstein's theories of relativity when he assumed the universe is coherently connected. According to the theories, we can never determine by measurement that the entire cosmos is causally connected at the same instant of time. Because of this consequence, it is not meaningful to say in physics, "The entire universe exists at one instant of time." See my discussion in Ronald C. Tobey, *The American Ideology of National Science, 1919–1930* (Pittsburgh: University of Pittsburgh Press, 1971), pp. 115–31.

75. For a brief summary of the history of "systems" in ecology, see Donald Worster, *Nature's Economy: The Roots of Ecology* (San Francisco: Sierra Club Books, 1977), pp. 40–41; also see A. J. Lotka, *The Elements of Mathematical Biology*, first edition, 1925; reprint edition (New York: Dover, 1956); Ludwig von Bertalanffy, *Modern Theories of Development, An Introduction to Theoretical Biology*, trans. and adapted, J. H. Woodger (London: Oxford University Press, 1933); Ludwig von Bertalanffy, *General System Theory* (New York: George Braziller, 1968). For an ideological, as well as historical analysis, see Robert Lilienfeld, The Rise of Systems Theory: An Ideological Analysis (New York: John Willy, 1978).

76. Tansley, "Use and Abuse," p. 305.

77. On the history of the concept of mechanical equilibrium systems, see Cynthia Eagle Russett, *The Concept of Equilibrium in American Social Thought* (New Haven: Yale University Press, 1966). On Bertrand Russell's theory of descriptions and names, see his famous article, "On Denoting" (1905).

78. In his correspondence with the Carnegie Institution, Clements clearly expressed his belief that he had accomplished the first step in the quantification of ecology; see F. E. Clements to Woodward, president of the Carnegie Institution, March 6, 1918, and Clements to Woodward, August 14, 1919, box 110, Clements Collection. R. A. Fisher, *The Genetical Theory of Natural Selection*, 1st ed., 1929; 2d. ed. rev. (New York: Dover, 1958).

79. Frederick Frost to Frederic Clements, January 14, 1925, box 113, Clements Collection.

80. Frederic Clements's comments, appended to Frost to Clements, January 14, 1925.

81. F. E. Clements and W. T. Penfound, "Experimental Evolution and Taxonomy," *Carnegie Institution of Washington Yearbook* 23 (1923–24):256–58; F. E. Clements, "Experimental Evolution," ibid. 24 (1924–25):309–320; F. E.

Clements, "Ecogenesis," ibid. 25 (1925–26):335–342; F. E. Clements, "Ecogenesis," ibid. 26 (1926–27):305–11. See also his comments on Frost to Clements, January 14, 1925. The classic disproof of these views was undertaken in a series of experiments in the 1930s and 1940's by Clausen and associates at Stanford and the University of California, Berkeley, and published in the four-volume study, Jens Clausen, David D. Keck, and William M. Hiesey, *Experimental Studies on the Nature of Species*, 4 vols. (Washington, D.C.: Carnegie Institution, 1940–58). On neolamarckism in American biology, see Edward J. Pfeifer, "The Genesis of American Neo-Lamarckism," *Isis* 56 (1965):156–67, and Hamilton Cravens, *The Triumph of Evolution: American Scientists and the Heredity-Environment Controversy, 1900–1941* (Philadelphia: University of Pennsylvania Press, 1978), pp. 34–39.

82. F. E. Clements to John Phillips, February 6, 1933, box 118, Clements Collection.

83. Tansley, "The Development of Vegetation."

84. Levy, *Universe of Science*, chap. 3, "The Queen of Sciences—Mathematics," pp. 82–117.

85. A. G. Tansley, [Review of Charles Elton, *Animal Ecology* (1927)], *Journal of Ecology* 16 (1928):167.

86. See George H. Shull to Frederic Clements, July 20, 1925, box 113; Frederic Clements to Frederic W. (unknown; Taylor?), April 30, 1942, box 124; Frederic Clements to Jeffrey (?), July 19, 1925, box 113; Frederic Clements to John C. Merriam, June 24, 1925, box 113, Clements Collection.

87. Tansley, "Use and Abuse," p. 299.

88. Samuel Haynes, *The Auden Generation: Literature and Politics in England in the 1930s* (New York: Viking Press, 1977); Gary Werskey, *The Visible College* (London: Allen Lane, 1978).

89. See n. 26.

90. Godwin, "Sir Arthur Tansley," p. 25.

91. On the Marxist (Stalinist) rejection of Freudian psychology, see V. N. Vološinov, *Freudianism: A Marxist Critique*, trans. I. R. Titunik, and ed. in collaboration with Neal H. Bruss (New York: Academic Press, 1976).

92. On Clements's sympathy for Tansley's educational fight, see Frederic Clements to A. G. Tansley, February 14, 1919, box 110, Clements Collection; Werskey, *Invisible College*, pp. 280–85.

93. A. G. Tansley, *Our Heritage of Wild Nature; A Plea for Organized Nature Conservation* (Cambridge: at the University Press, 1945), p. vii.

94. Thomas G. Manning, *Government in Science: The U.S. Geological Survey, 1867–1894* ([Lexington]: University of Kentucky Press, 1967), pp. 16–18, on the activities of F. V. Hayden; Rodernick Nash, *Wilderness and the American Mind* (New Haven and London: Yale University Press, 1967), chap. 7, "Wilderness Preserved," passim.

95. Tansley, *Our Heritage*, p. 1.

96. Ibid., p. vii.

97. Godwin, "Sir Arthur Tansley," p. 18.

98. Tansley, "Use and Abuse," p. 299.

99. Ibid., p. 297.

100. J. C. Smuts, *Holism and Evolution* (New York: Macmillan, 1926), pp. 35–58.

101. Ibid., pp. 315–16, 344.

102. Karl R. Popper, *The Open Society and Its Enemies*, revised edition (Princeton: Princeton University Press, 1950), pp. 225, 642n6. Popper was writing in the early years of World War II.

7. Saving the Prairies

1. John E. Weaver and F. W. Albertson, *Grasslands of the Great Plains* (Lincoln, Nebr.: Johnsen, 1956), pp. 76–78. Donald Worster, *Nature's Economy: The Roots of Ecology* (San Francisco: Sierra Club Books, 1977), pp. 221–53, deals with the impact of the drought on the Clementsian paradigm.

2. Weaver and Albertson, *Grasslands*, p. 83. Of course, these losses varied greatly from area to area. Ungrazed prairie, sheltered by a hill, might lose as little as 5 percent, while grazed prairie could lose that startling 95 percent. See ibid., chap. 5.

3. Ibid., p. 76.

4. John E. Weaver, *The Ecological Relations of Roots*, Carnegie Institution of Washington, publication 286 (Washington, D.C.: Carnegie Institution, 1919), pp. 1–2.

5. Ibid., p. 21.

6. Ibid., p. 118.

7. Ibid., p. 125.

8. Ibid.

9. Frederic E. Clements and John E. Weaver, *Experimental Vegetation: The Relation of Climaxes to Climates*, Carnegie Institution of Washington, publication 355 (Washington, D.C.: Carnegie Institution, 1924), pp. 5–6.

10. See the statement of their thesis in ibid., p. 7.

11. Ibid., p. 109.

12. Ibid., table 39, p. 111.

13. John E. Weaver and T. J. Fitzpatrick, "The Prairie," *Ecological Monographs* 4 (1934):281.

14. John E. Weaver and E. L. Flory, "The Stability of the Climax Prairie," *Ecology* 15 (1934):335.

15. Ibid., p. 336.

16. Ibid., p. 345.

17. Clements and Weaver, *Experimental Vegetation*, summary, pp. 142–43.

18. John E. Weaver and Frederic E. Clements, *Plant Ecology* (New York: McGraw-Hill, 1929), p. 43.

19. Ibid., pp. 43–44.

20. John E. Weaver and Frederic E. Clements, *Plant Ecology*, 2d ed. (New York: McGraw-Hill, 1938), p. 81.

21. Ibid., p. 272. A photograph of a billowing dust cloud threatening a Kansas town did appear on p. 248.

22. Ibid., p. 117. For a historical treatment of the thistle invasion, see Ronald Tobey, "Theoretical Science and Technology in American Ecology," *Technology and Culture* 17 (October, 1976):721–22.

23. Weaver and Clements, *Plant Ecology*, second edition, p. 272.

24. Ibid.

25. Ibid., p. 249.

26. Frederic E. Clements to A. G. Tansley, April 28, 1935, Dr. Edith G. Clements and Frederic E. Clements Collection, Division of Rare Books and Special Collections, The Library, University of Wyoming, Laramie, Wyoming; hereafter, Clements Collection.

27. Frederic E. Clements to John E. Weaver, December 27, 1934, Clements Collection; Frederic E. Clements and Ralph W. Chaney, *Environment and Life in the Great Plains*, Carnegie Institution of Washington, Supplementary Publications 24 (Washington, D.C.: Carnegie Institution, 1936), pp. 38–40.

28. J. E. Weaver and F. W. Albertson, "Effects of the Great Drought on the Prairies of Iowa, Nebraska, and Kansas," *Ecology* 17 (1936):570.

29. Ibid., p. 575.

30. Ibid., pp. 581–82.

31. Ibid., p. 586.

32. Ibid., p. 622.

33. J. E. Weaver and F. W. Albertson, "Deterioration of Midwestern Ranges," *Ecology* 21 (1940):228.

34. Ibid., p. 231.

35. Ibid., p. 235.

36. J. E. Weaver, "Replacement of True Prairie by Mixed Prairie in Eastern Nebraska and Kansas," *Ecology* 24 (1943):427.

37. Ibid., p. 420.

38. Ibid., p. 433.

39. J. E. Weaver, "The North American Prairie," *American Scholar* 13 (1944):339.

40. Karl Mannheim, *Ideology and Utopia; An Introduction to the Sociology of Knowledge*, trans. (from the German) Louis Wirth and Edward Shils, first edition, 1936 (New York: Harcourt, Brace and World, reprint edition, n.d.), p. 46.

41. Ibid., p. 79.

42. F. E. Clements to Walter Lowdermilk, June 9, 1934, Clements Collection.

43. F. E. Clements to Walter Lowdermilk, November 3, 1933; Walter Lowdermilk to F. E. Clements, May 10, 1935, Clements Collection. Edith Clements's memoir contains a colorful—if bizarre—account of her and Frederic's travels in dust bowl public service, but does not throw much light on her husband's intellectual development or ideological commitment; see Edith S. Clements, *Adventures in Ecology, Half a Million Miles . . . From Mud to Macadam* (New York: Pageant Press, 1960), pp. 220–44.

44. Jonathan Mitchell, "Shelter Belt Realities," *New Republic* 80 (August 29, 1934):69.

45. Ibid., pp. 69–71.

46. F. E. Clements to Bernhard [?], August 16, 1934, Clements Collection.

47. Walter Lowdermilk to F. E. Clements, August 12, 1934; F. E. Clements to Bernhard [?], August 16, 1934.

48. F. E. Clements to Walter Lowdermilk, December 24, 1934; F. E. Clements to Charles [?], December 24, 1934; F. E. Clements to Walter Lowdermilk, January 2, 1935, with the attached paper, "System of Exclosures [*sic*] for the Public Domain;" F. E. Clements to Walter Lowdermilk, January 19, 1935; F. E. Clements to Walter Lowdermilk, May 22, 1935, Clements Collection.

49. Clements, "System of Exclosures," pp. 2–3.

50. Ibid.

51. Raymond J. Pool, chairman, Department of Botany, to Chancellor E. A. Burnett, September 23, 1936, Chancellor Papers, University Archives, University of Nebraska, Lincoln, Nebraska.

52. F. E. Clements to Walter Lowdermilk, May 22, 1935, Clements Collection.

53. Quoted in F. E. Clements, "Experimental Ecology in the Public Service," *Ecology* 16 (1935):342.

54. On Huntington's problems in obtaining acceptance of his views and other aspects of his difficult career, see Geoffrey J. Martin, *Ellsworth Huntington, His Life and Thought* (Hamden, Conn.: Anchor Books, 1973).

55. Clements, "Experimental Ecology in the Public Service," p. 362.

56. Ibid., pp. 350–51.

57. Clements and Chaney, *Environment and Life*, p. 51.

58. Ibid.

59. Ibid., p. 52.

60. "The Week," *New Republic* 79 (August 8, 1934):329; Julian S. Bach, Jr., "Corn, Hogs, and Drought," ibid. 81 (December 5, 1934):98–99; Walter Clay Lowdermilk, *Palestine, Land of Promise* (New York and London: Harper & Brothers, 1944), chap. 11 on "The Jordan Valley Authority—A Counterpart of TVA in Palestine," and pp. 128–29 on cooperative settlement.

61. The Great Plains Committee, Morris L. Cooke, chairman, *The Future of the Great Plains*, House of Representatives, 75th Congress, 1st Session, Document No. 144, February 10, 1977, p. 2. On Cooke, see Kenneth E. Trombley, *The Life and Times of a Happy Liberal; A Biography of Morris Llewellyn Cooke* (New York: Harper, 1954).

62. Ibid., pp. 2–3.

63. Ibid., p. 11.

64. The committee's report cited an unpublished memorandum by Clements and also *Environment and Life in the Great Plains*; Great Plains Committee, *Future of the Great Plains*, pp. 32, 192.

65. Henry A. Wallace, *New Frontiers* (New York: Reynal and Hitch-

cock, 1934). On planning use of resources, see p. 22; on the necessity to control lumbering and grazing, see pp. 239—40, 245—46.

66. Ibid., p. 248.

67. See App. fig. 3, Absolute frequency of academic degrees.

68. See App. fig. 1.

69. See figs. 5, 6, 7.

70. Paul Sears, *Deserts on the March* (Norman: University of Oklahoma Press, 1935), *Lands Beyond the Forest* (Englewood Cliffs, N. J.: Prentice-Hall, 1969).

71. Thomas S. Kuhn, *The Structure of Scientific Revolutions*, 2d ed., enlarged (Chicago: University of Chicago Press, 1970), esp. chap. 12, "The Resolution of Revolutions," and chap. 14, "Progress Through Revolutions."

72. Diana Crane, *Invisible Colleges: Diffusion of Knowledge in Scientific Communities* (Chicago and London: University of Chicago Press, 1972), fig. 1, p. 172, and pp. 92, 104.

73. Wolfgang Stegmüller, *The Structure and Dynamics of Theories*, trans. (from the German) Dr. William Wohlhueter (New York, Heidelberg, Berlin: Springer-Verlag, 1976), chaps. 12 and 13, pp. 161—80, for a summary of the application of his theory to Kuhn's theory, and chaps. 12—16, pp. 161—231, passim, for discussion of his theory.

74. Edsko J. Dyksterhuis, "The Vegetation of the Fort Worth Prairie," *Ecological Monographs* 16 (1946): 12; publication of thesis of 1945.

In chronological order, the title of the theses follows, with publications that were carved out of them:

John William Crist, "Absorption of Nutrients from Subsoil in Relation to Crop Yield" (Ph.D. diss., University of Nebraska, Lincoln, 1923; dissertations hereafter cited with date alone), published with same title in *Botanical Gazette* 77 (April 1924);

Herbert C. Hanson, "A Study of the Vegetation of Northeastern Arizona" (1925), published with the same title in *University Studies*, University of Nebraska 24 (July—October 1924):85—178;

William Edward Bruner, "The Vegetation of Oklahoma" (1929), published with same title in *Ecological Monographs* 1 (1931):99—188;

Alfred E. Aldous, "Effect of Burning on Kansas Bluestem Pastures" (1934), published in part as "Bluestem Pastures," *Twenty-Eighth Biennial Report of the Kansas State Board of Agriculture* (1931—32), 33 (1933):184—91;

Laurence A. Stoddard, "Studies on Drought Resistance of Prairie Plants with Special Reference to Osmotic Pressure" (1934), published as "Osmotic Pressure and Water Content of Prairie Plants," *Plant Physiology* 10 (1935):561—680;

Fred William Albertson, "Ecology of the Mixed Prairie in West Central Kansas" (1937), published in part as "Prairie Studies in West Central Kansas," Kansas Academy of Sciences, *Transactions* 41 (1938):77—83;

Joseph Kramer, "Relative Efficiency of Roots and Taps of Plants in Protecting the Soil from Erosion" (1937);

Joseph Henry Robertson, "A Quantitative Study of True-Prairie After Three Years of Extreme Drought" (1939), with results partially published in J. E.

Weaver, Joseph H. Robertson, and Robert C. Fowler, "Changes in True-Prairie Vegetation During Drought as Determined by List Quadrats," *Ecology* 21 (1940):357–62;

Irene M. Mueller, "An Experimental Study of Rhizomes of Certain Prairie Plants" (1940), published with same title in *Ecological Monographs* 11 (1941): 165–188;

William Clarence Noll, "Environment and Physiological Activities of Winter Wheat and Prairie During a Year of Extreme Drought" (1943);

Edsko Jerry Dyksterhuis, "Vegetation of the Fort Worth Prairie" (1945), published under the same title in *Ecological Monographs* 16 (1946):1–29;

Raymond Winston Darland, "Relation of Height of Clipping or Grazing to Yield, Consumption, and Sustained Production of Certain Native Grasses" (1947);

Donald R. Cornelius, "Seed Production of Native Grasses Under Cultivation in Eastern Nebraska" (1949), published with the same title in *Ecological Monographs* 20 (1950):1–29; this thesis was conducted under the supervision of F. D. Keim, chairman of the Department of Agronomy, University of Nebraska, as well as Weaver;

Harold H. Hopkins, "Ecology of the Native Vegetation of the Loess Hills in Central Nebraska" (1949), published under the same title in *Ecological Monographs* 21 (1951):125–47;

John W. Voight, "Stages in Succession to True Prairie as Represented by Midwestern Pastures" (1950), published in part as "Range Condition Classes on Native Midwestern Pasture: An Ecological Analysis," *Ecological Monographs* 21 (1951):286–99;

Kling L. Anderson, "The Effect of Grazing Management and Site Conditions on Flint Hills Bluestem Pastures in Kansas" (1951), published in part, apparently, as "Utilization of Grasslands in the Flint Hills of Kansas," *Journal of Range Management* 6 (1953):86–93;

Gerald Wayne Tomanek, "Composition and Yield and Consumption in the Several Range Condition Classes in a Midwestern Range" (1951), did not, apparently, lead to a direct publication, although some of Tomanek's later publications were on tangential areas;

Farrel A. Branson, "Native Pastures of the Dissected Loess Plains of Central Nebraska" (1952).

75. Joseph F. Pechance, "Our Range Society," *Journal of Range Management* 1 (October 1948): 1–2; "Constitution and By-Laws of American Society of Range Management," ibid., p. 35.

76. "Officers, Council, Committees, and Members of American Society of Range Management, 1948," ibid., p. 40.

77. The significance of these events in creating the new profession was appreciated by Robert S. Campbell, "Milestones in Range Management," *Journal of Range Management*, ibid., pp. 4–8.

78. A membership list of the new society is printed in the first volume of its journal, ibid., pp. 40–62.

Methodological Appendix

1. On the history of the Ecological Society of America, see Robert L. Burgess, "The Ecological Society of America; Historical Data and Some Preliminary Analyses," in Frank N. Egerton, ed., *History of American Ecology* (New York: Arno Press, 1977).

2. On the constriction of the USDA budget in the 1920s, see W. M. Jardine [Secretary of Agriculture], "Report of the Secretary of Agriculture," United States Department of Agriculture, *Yearbook of Agriculture, 1928.* (Washington, D.C.: Government Printing Office, 1929) pp. 94–95.

3. Johnathan R. Cole and Stephen Cole, *Social Stratification in Science* (Chicago: University of Chicago Press, 1973).

4. Nonprofessional variables have been investigated especially in antebellum American science. See Clark Albert Elliott, "The American Scientist, 1800–1863: His Origins, Career, and Interests," Ph.D. diss., Case Western Reserve University, 1970, and Clark A. Elliott, "The American Scientist in Ante-bellum Society," *Social Studies of Science* 5 (1975):93–108. Also see Ronald C. Tobey, "American Grassland Ecology, 1895–1955: The Life-Cycle of a Professional Research Community," in Frank N. Egerton, ed., *History of American Ecology.*

5. Thomas Kuhn, *The Structure of Scientific Revolutions*, second edition, enlarged (Chicago: University of Chicago Press, 1970), p. ix. The classic citation analysis of the research front is D. K. de Solla Price, "Networks of Scientific Papers," *Science*, n.s. 149 (30 July 1965):510–15. On co-citation coupling, see Henry Small and Belver C. Griffith, "The Structure of Scientific Literatures, I: Identifying and Graphing Specialties," *Social Studies of Science* 4 (1974):17–40, and Griffith, Small, and Judith A. Stonehill and Sandra Dey, "The Structure of Scientific Specialties, II: Toward a Macrostructure and Microstructure for Science," ibid., pp. 339–65. Derek Price has corrected some of his early theories and provided a general mathematical theory of citation in "A General Theory of Bibliometric and Other Cumulative Advantage Processes," *Journal of the American Society for Information Science* 27 (1976):292–306. Critical reviews of citation analysis are provided by David Edge, "Quantitative Measures of Communication in Science: A Critical Review," *History of Science* 17 (1979):102–34, and Edward T. Morman, "Citation Analysis and the Current Debate over Quantitative Methods in the Social Studies of Science," 4s (1980) 3:7–13.

6. H. Gilman McCann, *Chemistry Transformed: The Paradigmatic Shift from Phlogiston to Oxygen* (Norwood, N.J.: Ablex, 1978).

7. Ibid., chap. 5, "Models of Revolution."

8. See the references in n. 5.

9. Price, "Networks of Scientific Papers," pp. 511–12.

BIBLIOGRAPHY

Manuscript Sources

Alumni Records, Alumni Association, University of Nebraska, Lincoln, Nebraska.

Chancellor Papers, University Archives, University of Nebraska, Lincoln, Nebraska.

Charles E. Bessey Papers, University Archives, University of Nebraska, Lincoln, Nebraska.

Dr. Edith G. Clements and Frederic E. Clements Collection, Division of Rare Books and Special Collections, The Library, University of Wyoming, Laramie, Wyoming.

Edith Gertrude Clements Papers, Nebraska State Historical Society, State Archives Division, Manuscript Division, Lincoln, Nebraska.

Nathan Roscoe Pound Papers, Nebraska State Historical Society, State Archives Division, Manuscript Division, Lincoln, Nebraska.

Per Axel Rydberg Papers, Nebraska State Historical Society, State Archives Division, Manuscript Division, Lincoln, Nebraska.

Records of the Sem. Bot. Club, University Archives, University of Nebraska, Lincoln, Nebraska.

Registration Records, Registrar, University of Nebraska, Lincoln, Nebraska.

Unpublished Sources

Elliott, Clark Albert. "The American Scientist, 1800–1863: His Origins, Career, and Interests." Ph.D. dissertation. Case Western Reserve University, 1970.

Liebetrau, Suzayne Fries. "Trailblazers in Ecology—The American Ecology Consciousness." Ph.D. dissertation. University of Michigan, 1973.

Walsh, Thomas R. "Charles E. Bessey: Land-Grant Professor." Ph.D. dissertation. University of Nebraska, 1972.

Webber, Herbert John. "On the Trail of the Orange: The Autobiography of an Ordinary Man." Dictated, 1945. Biological Sciences Library, University of California, Riverside.

Published Sources

Prefatory Note

I have not listed the full scientific literature of grassland ecology that is referred to in chapter 5. A computer printout of all 535 titles is available by writing to the author. The fifty-four titles most frequently cited in the grassland literature, which are the basis of the generational analysis in chapter six, are listed with full bibliographical information in appendix tables 12, 13, 14, 15, and 16, and are not repeated here. Finally, I have not listed all the important titles in the history of ecology during the period that related to the Clementsian paradigm; these titles are critically discussed in the monographs by Egerton (1976), Phillips (1934, 1935), Whittaker (1962), and McIntosh (1967, 1976), which are referenced below. As a consequence, most entries in this bibliography are "secondary" sources, or listed for their historical information.

Adams, C. C. "The Ecological Succession of Birds." *Annual Report, Michigan Geological Survey* (1908): 121–54.

————. *Guide to the Study of Animal Ecology*. New York: Macmillan, 1913.

Adams, Frank Dawson. *The Birth and Development of the Geological Sciences*. Reprint edition. New York: Dover, 1954.

Allen, David Elliston. *The Naturalist in Britain: A Social History*. London: Allen Lane, 1976.

Allen, Walter. *Urgent West: The American Dream and Modern Man*. New York: Dutton, 1969.

Allee, W. C.; Emerson, Alfred; Park, Orlando; and Schmidt, Karl P. *Principles of Animal Ecology*. Philadelphia: Saunders, 1949.

Arthur, J. C. "Development of Vegetable Physiology." *Botanical Gazette* 20 (1895): 381–402.

————. "Some Botanical Laboratories of the United States." *Botanical Gazette* 10 (1885):395–406.

Bailey, L. H. *The Country-Life Movement in the United States*. New York: Macmillan, 1913.

Baker, Gladys L., et al. *Century of Service: The First 100 Years of the United States Department of Agriculture*. Washington, D.C.: Centennial Committee, U.S. Department of Agriculture, 1963.

Beck, Hanno. *Alexander von Humboldt*. 2 vols. Wiesbaden: Franz Steiner Verlag GMBH, 1959–61.

Ben-David, Joseph. *The Scientist's Role in Society: A Comparative Study*. Englewood Cliffs, N.J.: Prentice-Hall, 1971.

Bennett, Mildred R. *The World of Willa Cather*. Reprint edition. Lincoln: University of Nebraska Press, 1961.

Bertalanffy, Ludwig von. *General System Theory*. New York: George Braziller, 1968.

————. *Modern Theories of Development: An Introduction to Theoretical Biology*, translated and adapted by J. H. Woodger. London: Oxford University Press, 1933.

Bessey, C. E. "Botany by the Experimental Method." *Science* n.s. 35 (1912):994–96.
_____. *Botany for High Schools and Colleges.* New York: Henry Holt, 1880.
_____. *The Essentials of Botany.* Fifth edition. New York: Henry Holt, 1893.
_____. "High School Botany." *Science* n.s. 7 (1898):266–67.
_____. "On the Preparation of Botanical Teachers." *Science* n.s. 33 (1911):633–39.
Bessey, Ernst A. "The Teaching of Botany Seventy-five Years Ago." *Iowa State College Journal of Science* 9 (1935):13–19.
Blackman, F. F. and Tansley, A. G. "Ecology in Its Physiological and Phyto-geographical Aspects." *The New Phytologist* 4 (1905):90–203, 232–53.
Boller, Paul F., Jr. *American Thought in Transition: The Impact of Evolutionary Naturalism, 1865–1900.* Chicago: Rand McNally, 1969.
Bowers, William L. *The Country Life Movement in America, 1900–1920.* Port Washington, N.Y.: Kennikat Press, 1974.
Braun, E. Lucy. "The Development of Association and Climax Concepts, Their Use in Interpretation of the Deciduous Forest." *Fifty Years of Botany,* edited by William Campbell Steere. New York: McGraw-Hill, 1958.
Brewer, Richard. "A Brief History of Ecology; Part I—Pre-Nineteenth Century to 1919." *Occasional Papers of the C. C. Adams Center for Ecological Studies,* no. 1. Kalamazoo, Mich., 1960.
Brown, E. K. *Willa Cather, A Critical Biography.* Completed by Leon Edel. New York: Knopf, 1953.
Browning, Douglas, ed. *Philosophers of Process.* New York: Random House, 1965.
Bruhns, Karl C., ed. *Alexander von Humboldt.* 3 vols. Leipzig:Brockhaus, 1872.
_____. *Life of Alexander von Humboldt.* 2 vols. Trans. Jane and Caroline Lassell. London: Longmans, Green, 1873.
Cameron, Jenks. *The Development of Governmental Forest Control in the United States.* Baltimore: Johns Hopkins University Press, 1928.
De Candolle, Alphonse. *Géographie botanique raisonnée; ou exposition des faits principaux et des lois concernant la distribution géographique des plantes de l'époque actulle.* 2 vols. Paris: Librairie de Victor Masson, 1855.
Carlson, Paul H. "Forest Conservation on the South Dakota Prairies." *South Dakota History* 2 (1971):23–45.
Chisholm, Anne. *Philosophers of the Earth, Conversations with Ecologists.* New York: Dutton, 1972.
Clarke, Robert. *Ellen Swallow: The Woman Who Founded Ecology.* Chicago: Follett, 1973.
Clements, Edith S. *Adventures in Ecology, Half a Million Miles . . . From Mud to Macadam.* New York: Pageant Press, 1960.
Clements, Frederic E. "Darwin's Influence Upon Plant Geography and Ecology." *American Naturalist* 43 (1909):143–51.
_____. *The Development and Structure of Vegetation.* University of Nebraska, Botanical Survey of Nebraska, 7, Studies in the Vegetation of the State, 3. Lincoln, Nebr.: Published by the Seminar, 1904.
_____. "The Ecological Method in Teaching Botany." *New Phytologist* 22 (1923):98–104.
_____. "Experimental Ecology in the Public Service." *Ecology* 16 (1935):342–63.

———. "Formation and Succession Herberia." University Studies. *University of Nebraska* 4 (1904):329–55.

———. "Nature and Structure of the Climax." *Journal of Ecology* 24 (1936):252–84.

———. *Plant Physiology and Ecology.* New York: Henry Holt, 1907.

———. "Methods of Botanical Teaching." *Science* n.s. 33 (1911):642–46.

———, and Chaney, Ralph W. *Environment and Life in the Great Plains.* Washington, D.C.: Carnegie Institution, 1936.

———, and Cutter, Irving S. *Laboratory Exercises in General Botany.* Lincoln, Nebr.: University Publishing Co., 1900.

———, and Goldsmith, Glenn W. *The Phytometer Method in Ecology: The Plant and Community as Instruments.* Washington, D.C.: Carnegie Institution, 1924.

———, and Shelford, Victor E. *Bio-Ecology.* New York: John Wiley, 1939.

Clouser, Roger A. *Man's Intervention in the Post-Wisconsin Vegetational Succession of the Great Plains.* Occasional Paper no. 4. Department of Geography-Meteorology, University of Kansas, 1978.

Coleman, William. *Biology in the Nineteenth Century: Problems of Form, Function, and Transformation.* New York: John Wiley, 1971.

Collins, H. M. "The TEA Set: Tacit Knowledge and Scientific Networks." *Social Studies of Science* 4 (1974):165–85.

Conrad, H. S. *The Background of Plant Ecology.* Translation from the German of A. Kerner, *The Plant Life of the Danube Basin.* Ames: Iowa State College Press, 1951.

———. "Plant Associations on Land." *American Midland Naturalist* 21 (1939):1–26.

Cooke, Morris L., chairman, The Great Plains Committee. *The Future of the Great Plains.* Washington, D.C.: U.S. Government Printing Office, 1937.

Costello, David F. *The Prairie World.* New York: Thomas Y. Crowell, 1969.

Coulter, John Merle. "Articles on Laboratories, Appliances, and Courses of Instruction." *Botanical Gazette* 10 (1885):409–13, 417–21.

———. "Botanical Teaching." *Science* n.s. 33 (1911):646–49.

———. "Botany as a Factor in Education." *School Review* 12 (1904):609–17.

———. *Plant Relations: A First Book of Botany.* New York: D. Appleton, 1900.

———, Barnes, C. R., and Cowles, H. C. *A Textbook of Botany for Colleges and Universities.* 2 vols. New York: American Book Co., 1911.

———, and Coulter, Merle C. *Where Evolution and Religion Meet.* New York: Macmillan, 1924.

Cowles, H. C. "The Causes of Vegetative Cycles." *Botanical Gazette* 51 (1911): 161–83.

———. "The Ecological Relations of the Vegetation on the Sand Dunes of Lake Michigan." *Botanical Gazette* 27 (1899):95–117, 167–202, 281–308, 361–91.

———. "The Influence of Underlying Rocks on the Character of Vegetation." *Bulletin,* American Bureau of Geography 2 (1901):163–76, 376–88.

———. Review of *Plant Succession. Botanical Gazette* 68 (1919):477–78.

Crane, Diana. *Invisible Colleges: Diffusion of Knowledge in Scientific Communities.* Chicago and London: University of Chicago Press, 1972.

Cravens, Hamilton. *The Triumph of Evolution: American Scientists and the Heredity-Environment Controversy, 1900–1941*. Philadelphia: University of Pennsylvania Press, 1978.

Dachnowski, Alfred. "The International Phytogeographical Excursion of 1913 and Its Significance to Ecology in America." *Journal of Ecology* 2 (1914): 237–45.

Dana, Samuel Trask. *Forest and Range Policy: Its Development in the United States*. New York: McGraw-Hill, 1956.

Daniels, George H. *American Science in the Age of Jackson*. New York and London: Columbia University Press, 1968.

———. *Science in American Society, A Social History*. New York: Knopf, 1971.

Darwin, Charles. *The Origin of Species by Means of Natural Selection*. Facsimile edition of the first edition, edited by Ernst Mayr. New York: Athaneum, 1967.

———. *The Origin of Species by Means of Natural Selection*. Sixth edition. London: John Murray, 1900.

Davenport, Charles B. *Statistical Methods, with Special Reference to Biological Variation*. New York: John Wiley, 1899.

Davies, Gordon L. *The Earth in Decay: A History of British Geomorphology, 1578–1878*. London: Macdonald Technical and Scientific, n.d.

Dodds, Gordon B., ed. "Conservation and Reclamation in the TransMississippi West: A Critical Bibliography." *Arizona and the West* 13 (1971):143–71.

Drude, Oscar. *Deutschlands Pflanzengeographie: Ein geographes Charakterbild der Flora von Deutschland*. Stuttgart: Verlag von J. Engelhorn, 1896.

———. *Handbuch der Pflanzengeographie*. Stuttgart: Verlag von J. Engelhorn, 1890.

Dupree, A. Hunter. *Asa Gray, 1810–1888*. Reprint edition. New York: Atheneum, 1968.

———. *Science in the Federal Government: A History of Policies and Activities to 1940*. Reprint edition. New York: Harper & Row, Harper Torchbooks, 1957.

Edge, David O., and Mulkay, Michael J. *Astronomy Transformed: The Emergence of Radio Astronomy in Britain*. New York: John Wiley, 1976.

Edwards, Everett E. *A Bibliography of the History of Agriculture in the United States*. Washington, D.C.: U.S. Department of Agriculture, 1930.

Egerton, Frank N. "A Bibliographical Guide to the History of General Ecology." *History of Science* 15 (1977):189–215.

———. "Changing Concepts of the Balance of Nature." *The Quarterly Review of Biology* 48 (1973):322–50.

———. "Ecological Studies and Observations Before 1900." In *Issues and Ideas in America*, edited by Benjamin J. Taylor and Thurman J. White. Norman: University of Oklahoma Press, 1976.

———, ed. *History of American Ecology*. New York: Arno Press, 1977.

———, editor. *History of Ecology*. 52 vols. An Arno Press Collection. Reprint. New York: Arno Press, 1977.

———. "Humboldt, Darwin, and Population." *Journal of the History of Biology* 3 (1970):325–60.

Egleston, N. H. *Arbor Day: Its History and Observance*. Washington: U.S.

Department of Agriculture, 1896.

Ekirch, Arthur A., Jr. *Man and Nature in America*. New York and London: Columbia University Press, 1963.

Elliott, Clark A. "The American Scientist in Antebellum Society." *Social Studies of Science* 5 (1975):93–108.

Elton, Charles. *Animal Ecology*. New York: Macmillan, 1927.

Ewan, Joseph, "Bessey, Charles Edwin." *Dictionary of Scientific Biography*, edited by Charles Coulton Gillispie. New York: Charles Scribner's Sons, 1970.

Farlow, William G. "The Change from the Old to the New Botany in the United States." *Science* n.s. 37 (1913):79–86.

Fein, Albert. *Frederick Law Olmsted and the American Environmental Tradition*. New York: George Braziller, 1972.

Feuer, Lewis. *Einstein and the Conflict of Generations*. New York: Basic Books, 1974.

Flader, Susan L. *Thinking Like a Mountain; Aldo Leopold and the Evolution of an Ecological Attitude Toward Deer, Wolves, and Forests*. Columbia: University of Missouri Press, 1974.

Fleming, Donald. "Roots of the New Conservation Movement." *Perspectives in American History* 6 (1972):7–94.

Ford, Charles E. "Botany Texts: A Survey of Their Development in American Higher Education, 1643–1906." *History of Education Quarterly* 4 (1964):59–70.

Ganong, W. "Suggestions for an Attempt to Secure a Standard College Entrance Option in Botany." *Science* n.s. 13 (1901):611–16.

Ghiselin, Michael T. *The Economy of Nature and the Evolution of Sex*. Berkeley, Los Angeles, London: University of California Press, 1974.

Glacken, Clarence J. *Traces on the Rhodian Shore: Nature and Culture in Western Thought from Ancient Times to the End of the Eighteenth Century*. Reprint edition. Berkeley, Los Angeles, London: University of California Press, 1976.

Gleason, H. A. "The Individualistic Concept of the Plant Association." *American Midland Naturalist* 21 (1939):92–110.

———. "The Structure and Development of the Plant Association." *Bulletin of the Torrey Botanical Club* 44 (1917):463–81.

———. "Twenty-five Years of Ecology, 1910–1935." In *Twenty-Five Years of Progress in Botany, 1910–1935*, edited by C. S. Gager. *Brooklyn Botanical Garden Memoirs* 4 (1936).

Godwin, Sir Harry. "Arthur George Tansley, 1871–1955." *Biographical Memoirs of Fellows of the Royal Society* 3 (1957):227–46.

———. "Sir Arthur Tansley: The Man and the Subject." *The Journal of Ecology* 65 (1977):1–26.

Goethe, Johann W. Von. *The Metamorphosis of Plants*. In *Goethe's Botany: The Metamorphosis of Plants (1790) and Tobler's Ode to Nature (1782)*, translated by Agnes Arber. Waltham, Mass.: Chronica Botanica, 1946.

Golley, Frank B., ed. *Ecological Succession*. Benchmark Papers in Ecology 15. Stroudsburgh, Pennsylvania: Dowden, Hutchinson & Ross, 1977.

Gould, Stephen Jay. *Ontogeny and Phylogeny*. Cambridge, Mass.: Belknap Press of Harvard University Press, 1977.

Green, Joseph Reynolds. *A History of Botany, 1860–1900: Being a Continuation of Sachs History of Botany, 1530–1860.* New York: Russell & Russell, 1967.

———. *A History of Botany in the United Kingdom from the Earliest Times to the End of the 19th Century.* London: J. M. Dent, 1914.

Greene, Edward L. *Pittonia: A Series of Papers Relating to Botany and Botanists,* 1–5 (1887–1905).

Grisebach, August. *Gesammelte Abhandlungen und kleinere Schriften zur Pflanzengeographie.* Liepzig: Verlag von Wilhelm Engelmann, 1880.

———. "Pflanzengeographie und Botanik." In *Alexander von Humboldt: Eine wissenschaftliche Biographie,* edited by Karl C. Bruhns. 3 vols. Leipzig: Brockhaus, 1872.

Guralnick, Stanley M. *Science and the Ante-bellum American College.* Memoirs of American Philosophical Society, 109. Philadelphia: American Philosophical Society, 1975.

Hagstrom, Warren. *The Scientific Community.* New York: Basic Books, 1965.

Hankins, Frank H. "Adolphe Quetelet as Statistician." *Studies in History, Economics, and Public Law* 31 (1908):445–576.

Hanley, Wayne. *Natural History in America, From Mark Catesby to Rachel Carson.* New York: A Demeter Press Book, Quadrangle/New York Times, 1977.

Hanson, Herbert C., and Churchill, Ethan D. *The Plant Community.* New York: Reinhold Publishing Corp. 1961.

Hawkins, Hugh. *Between Harvard and America: The Educational Leadership of Charles W. Eliot.* New York: Oxford University Press, 1972.

Haynes, Samuel. *The Auden Generation: Literature and Politics in England in the 1930s.* New York: Viking Press, 1977.

Hendrickson, Walter B. "Science and Culture in Nineteenth Century Michigan." *Michigan History* 57 (1973):140–50.

Higham, John, and Conkin, Paul K., eds. *New Directions in American Intellectual History.* Baltimore and London: Johns Hopkins University Press, 1979.

Holton, Gerald. *Thematic Origins of Scientific Thought, Kepler to Einstein.* Reprint Edition. Cambridge, Mass.: Harvard University Press, 1975.

Horowitz, Helen Lefkowitz. *Culture and the City: Cultural Philanthropy in Chicago from the 1880s to 1917.* Lexington: University of Kentucky Press, 1976.

Hoskins, William G. *The Making of the English Landscape.* London: Hodder and Stoughton, 1955.

Humboldt, Alexander von. *Aspects of Nature,* translated from the German by Mrs. Sabine. Third edition; first English edition. 2 vols. London: Longman, Brown, Green, and Longmans, 1849.

———. *Cosmos, A Sketch of the Universe,* translated by E. C. Otto. 2 vols. New York: Harper, 1850.

———, and Bonpland, Aimé. *Essai sur la géographie des plantes.* Paris: Libraire Lebrault Schoell, 1807.

Huth, Hans. *Nature and the American: Three Centuries of Changing Attitudes.* Berkeley: University of California Press, 1957.

Jones, Howard Mumford. *The Age of Energy: Varieties of American Experience, 1865–1915.* Reprint edition. New York: Viking Press, 1973.

Kant, Immanuel. *Critique of Judgment*. In *The Philosophy of Kant; Immanuel Kant's Moral and Political Writings*, edited by Carl J. Friedrich, translated by James C. Meredith. New York: Modern Library, 1949.

Kendeigh, S. C. "History and Evaluation of Various Concepts of Plant and Animal Communities in North America." *Ecology* 35 (1954):152–71.

Kirschner, Don S. *City and Country: Rural Responses to Urbanization in the 1920s*. Westport, Conn.: Westport Publishing, 1970.

Kitchen, Paddy. *A Most Unsettling Person: The Life and Ideas of Patrick Geddes, Founding Father of City Planning and Environmentalism*. New York: Saturday Review Press, 1976.

Kohlstedt, Sally Gregory. *The Formation of the American Scientific Community: The American Association for the Advancement of Science, 1848–1860*. Urbana: University of Illinois Press, 1976.

Krug, Edward A. *The Shaping of the American High School, 1880–1920*. Madison, Milwaukee, and London: University of Wisconsin Press, 1960.

Kuhn, Thomas. *The Copernican Revolution, Planetary Astronomy in the Development of Western Thought*. Reprint Edition. New York: Vintage Books, 1959.

———. *The Structure of Scientific Revolutions*. Enlarged, second edition. Chicago: University of Chicago Press, 1970.

Laudan, Larry. *Progress and Its Problems: Toward a Theory of Scientific Growth*. Berkeley, Los Angeles, London: University of California Press, 1977.

Lemaine, Gerard, et al., eds. *Perspectives on the Emergence of Scientific Disciplines*. Paris and The Hague: Mouton; Chicago: Aldine, 1976.

Levy, H. *The Universe of Science*. London: Watts, 1938.

Lilienfeld, Robert. *The Rise of Systems Theory: An Ideological Analysis*. New York: John Wiley, 1978.

Lotka, A. J. *The Elements of Mathematical Biology*. Reprint edition. New York: Dover, 1956.

Lovejoy, Arthur O. *The Great Chain of Being: A Study in the History of Ideas*. Reprint edition. Cambridge, Mass.: Harvard University Press, 1964.

Lowdermilk, Walter Clay. *Conquest of the Land Through Seven Thousand Years*. Revised edition. Washington, D.C.: U.S. Government Printing Office, 1950.

———. "Lessons from the Old World to Americas in Land Use." In *Annual Report, Smithsonian Institution, 1943*. Washington, D.C.: Smithsonian Institution. 1944.

———. *Palestine, Land of Promise*. New York and London: Harper & Brothers, 1944.

———. *Soil Erosion and Its Control in the United States*. Washington, D.C.: Soil Conservation Service, 1935.

Lowe, P. D. "Amateurs and Professionals: The Institutional Emergence of British Plant Ecology." *Journal of the Society for Bibliography of Natural History* 7 (1976):517–35.

Lurie, Edward. *Louis Agassiz, A Life in Science*. Abridged reprint edition. Chicago and London: University of Chicago Press, Phoenix Books, 1960.

Lyell, Charles. *Elements of Geology*. London: John Murray, 1838.

———. *Principles of Geology*. 2 vols. London: John Murray, 1830–32.

Lynch, Kevin. *The Image of the City*. Cambridge, Mass.: Technology Press, 1960.

McCann, H. Gilman. *Chemistry Transformed: The Paradigmatic Shift from Phlogiston to Oxygen*. Norwood, N. J.: Ablex, 1978.

McHenry, Robert. *A Documentary History of Conservation in America*. New York: Praeger, 1972.

McIntosh, Robert P. "The Continuum Concept of Vegetation." *The Botanical Review* 33 (1967): 130−87.

———. "Ecology Since 1900." In *Issues and Ideas in America*, edited by Benjamin J. Taylor and Thurman J. White. Norman: University of Oklahoma Press, 1976.

———. "H. A. Gleason—'Individualistic Ecologist' 1882−1975: His Contributions to Ecology." *Bulletin of the Torrey Botanical Club* 102 (1975): 253−73.

———. "Plant Ecology, 1947−1972." *Annals of the Missouri Botanical Garden* 61 (1974):132−65.

Magnus, Rudolf. *Goethe as a Scientist*, translated by Heinz Norden. New York: Henry Schuman, 1949.

Major, Jack. "Historical Development of the Ecosystem Concept." Pp. 9−22 in George M. Van Dyne, *The Ecosystem Concept in Natural Resource Management*. New York: Academic Press, 1969.

Manley, Robert N. *Centennial History of the University of Nebraska; I. Frontier University (1869−1919)*. Lincoln: University of Nebraska Press, 1969.

Mannheim, Karl. *Ideology and Utopia: An Introduction to the Sociology of Knowledge*, translated from the German by Louis Wirth and Edward Shils. Reprint edition. New York: Harcourt, Brace & World, a Harvest Book, n.d.

Martin, Geoffrey J. Martin. *Ellsworth Huntington, His Life and Thought*. Hamden, Conn.: Anchor Books, 1973.

May, Henry F. *The End of American Innocence: A Study of the First Years of Our Own Time, 1912−1917*. Reprint edition. Chicago: Quadrangle Books, 1964.

Meitzen, August. *History—Theory and Technique of Statistics*. Translated by Roland P. Falkner. *Supplement to the Annals of the American Academy of Political and Social Science* (March, 1891).

Miller, Howard S. *Dollars for Research: Science and its Patrons in Nineteenth-Century America*. Seattle and London: University of Washington Press, 1970.

Miner, J. R. "Pierre-Francois Verhulst, the Discoverer of the Logistic Curve." *Human Biology* 5 (1933):673−89.

Morgan, C. Lloyd. *Emergent Evolution*. London: Williams and Norgate, 1923.

———. *The Interpretation of Nature*. Bristol, England: J. W. Arrowsmith, 1905.

Mott, Frank L. *A History of American Magazines*. 5 vols. Cambridge, Mass.: Harvard University Press, 1957−68.

Müller, D. "Warming, Johannes Eugenius Bülow." *Dictionary of Scientific Biography*, edited by Charles Coulston Gillispie. New York: Charles Scribner's Sons, 1976.

Murphy, William Michael, and Bruckner, D. J. R., eds. *The Idea of the University of Chicago: Selections from the Papers of the First Eight Chief Executives of the University of Chicago, 1891−1975*. Chicago and London: University of Chicago Press, 1976.

Myer, D. H. *The Instructed Conscience: The Shaping of the American National Ethic.* Philadelphia: University of Pennsylvania Press, 1972.

Nash, Roderick. *Wilderness and the American Mind.* New Haven and London: Yale University Press, 1967.

Newcombe, F. C. "Equipment and Administration of the High-School Botanical Laboratory." *School Review* 7 (1899):301—8.

Nisbet, H. B. *Goethe and the Scientific Tradition.* London: Institute of Germanic Studies, 1972.

Nordenskijöld, Erik. *The History of Biology, A General Survey.* New York: Tudor Publishing, 1928.

Oparin, A. I. *The Origin of Life*, translated by Serguij Morgulis. First edition, 1938. Reprint edition. New York: Dover, 1953.

Overfield, Richard A. "Charles E. Bessey: The Impact of the New Botany on American Agriculture." *Technology and Culture* 16 (1975):162—81.

———. "Trees for the Great Plains: Charles E. Bessey and Forestry." *Journal of Forest History* 23 (January, 1979):18—31.

Pammel, L. J. "Dr. Charles Edwin Bessey." In *Prominent Men I Have Met.* Ames: Iowa State College, 1925.

Pearson, E. S. *Karl Pearson: An Appreciation of Some Aspects of His Life and Work.* New York: Macmillan, 1938.

———, and Kendall, M. G. *Studies in the History of Statistics and Probability.* London: Griffin, 1970.

Pearson, Karl. *The Life, Letters and Labors of Francis Galton.* 3 vols. Cambridge [England]: Cambridge University Press, 1914—30.

Peirce, Charles Saunders. *Philosophical Writings*, edited by Justus Buchler. Reprint edition. New York: Dover, 1955.

Persons, Stow. *The Decline of American Gentility.* New York and London: Columbia University Press, 1973.

Phillips, D. C. "Organicism in The Late Nineteenth and Early Twentieth Centuries." *Journal of the History of Ideas* 31 (1970): 413—32.

Phillips, John. "The Biotic Community." *Journal of Ecology* 19 (1931):1—24.

———. "Succession, Development, the Climax, and the Complex Organism: An Analysis of the Concepts." *Journal of Ecology* 22 (1934):554—71; 23 (1935):210—46, 488—508.

Polanyi, Michael. *Personal Knowledge: Towards a Post-Critical Philosophy.* Chicago: The University of Chicago Press, 1958.

———. *The Tacit Dimension.* Garden City, New York: Doubleday, 1966.

Pool, Raymond J. "Evolution and Differentiation of Laboratory Teaching in the Botanical Sciences." *Symposia Commemorating Six Decades of Botanical Science, Iowa State College Journal of Science* 9 (1934—1935):235—42.

Popper, Karl R. *The Open Society and Its Enemies.* Revised edition. Princeton: Princeton University Press, 1950.

Porter, Roy. *The Making of Geology: Earth Science in Britain, 1660—1815.* Cambridge [England]: Cambridge University Press, 1977.

Pound, Roscoe, and Clements, Frederic E. "A Method of Determining the Abundance of Secondary Species." *Minnesota Botanical Studies* 2 (1898): 19—24.

————, and ————. "Vegetational Regions of the Prairie Province." *Botanical Gazette* 25 (1898):382–83.

Price, Derek de Solla. "A General Theory of Bibliometric and Other Cumulative Advantage Processes." *Journal of the American Society for Informational Science* 27 (1976):292–306.

————. "Networks of Scientific Papers." *Science* 149 (1965):510–15.

Pringsheim, Ernst G. *Julius Sachs, der Begründer der neueren Pflanzenphysiologie, 1832–1897.* Jena: Fischer, 1932.

Ramaley, Francis. "The Growth of a Science." *University of Colorado Studies* 26 (1940):3–14.

Reinke, J. "Grisebach, August." *Botanische Zeitung* 37 (1879):521–34.

Rodgers, Andrew Denny, III. *American Botany, 1873–1892: Decades of Transition.* Facsimile of the edition of 1944. New York and London: Hafner, 1968.

————. *John Merle Coulter, Missionary in Science.* Princeton: Princeton University Press, 1944.

Rollins, Peter, and Elder, Harris J. "Environmental History in Two New Deal Documentaries." *Film and History* 3 (1973):1–7.

Rosenberg, Charles E. "Science, Technology, and Economic Growth: The Case of the Agricultural Experiment Station Scientist, 1875–1914." *Agricultural History* 45 (1971):1–20.

Rubel, E. "Ecology, Plant Geography and Geo-botany: Their History and Aim." *Botanical Gazette* 84 (1927):428–39.

Rudolph, Frederick. *The American College and University, A History.* New York: Knopf, 1962.

Russett, Cynthia E. *The Concept of Equilibrium in American Social Thought.* New Haven: Yale University Press, 1966.

Sachs, Julius. *History of Botany (1530–1860),* translated by Henry E. F. Garnsey. Oxford: Clarendon Press, 1906.

————. *Text-book of Botany, Morphological and Physiological,* translated by Alfred W. Bennett. Oxford: Clarendon Press, 1875.

Saloutos, Theodore. "The New Deal and Farm Policy in the Great Plains." *Agricultural History* 43 (1969):345–55.

Santayana, George. "The Genteel Tradition in American Philosophy." In *Winds of Doctrine and Platonism and the Spiritual Life.* Reprint edition. New York: Harper & Brothers, Harper Torchbooks, 1957.

Schmidt, Peter. *Back to Nature: Arcadian Myth in Urban America.* New York: Oxford University Press, 1969.

Schneer, Cecil. "The Rise of Historical Geology in the Seventeenth Century." *Isis* 45 (1954):256–68.

Schneider, Herbert W. *A History of American Philosophy.* Reprint edition. New York and London: Columbia University Press, 1963.

Sears, Paul B. "Botanists and the Conservation of Natural Resources." In *Fifty Years of Botany,* edited by William Campbell Steere. New York: McGraw-Hill, 1958.

————. "Clements, Frederic Edward." *Dictionary of American Biography,* edited by Edward T. James. New York: Charles Scribner's Sons, 1973.

————. "Coulter, John Merle." *Dictionary of American Biography, Supplement.*

———. *Deserts on the March.* Norman: University of Oklahoma Press, 1935.

———. *Lands Beyond the Forest.* Englewood Cliffs, N. J.: Prentice-Hall, 1969.

———. *The Living Landscape.* New York: Basic Books, 1966.

———. "Plant Ecology." In *A Short History of Botany in the United States*, edited by Joseph Ewan. New York and London: Hafner, 1969.

"Shaw School of Botany," *Science* n.s. 29 (1909):693–94.

Shepard, Paul. *Man in the Landscape: A Historic View of the Esthetics of Nature.* New York: Knopf, 1967.

Sizer, Theodore R. *Secondary Schools at the Turn of the Century.* New Haven and London: Yale University Press, 1964.

Shimwell, David W. *The Description and Classification of Vegetation.* Seattle: University of Washington Press, 1971.

Small, Henry, and Griffith, Belver C. "The Structure of Scientific Literatures, I: Identifying and Graphing Specialties." *Social Studies of Science* 4 (1974): 17–40.

———, ———, Stonehill, Judith A., and Dey, Sandra. "The Structure of Scientific Specialties, II: Toward a Macrostructure and Microstructure for Science." *Social Studies of Science* 4 (1974):339–65.

Smallwood, W. J., and Smallwood, Mabel. *Natural History and the American Mind.* New York: Columbia University Press, 1941.

Smith, Henry Nash. *Virgin Land: The American West as Symbol and Myth.* Reprint edition. New York: Vintage Books, n.d., first edition, 1950.

Smuts, J. C. *Holism and Evolution.* New York: Macmillan, 1926.

Spaulding, V. M. "Rise and Progress of Ecology." *Science* n.s. 17 (1903):201–10.

Sperry, Theodore M. "The Kansas Prairie: Pre-History of Crawford County." *Midwest Quarterly* 12 (1971):271–87.

Stauffer, Robert C. "Ecology in the Long Manuscript Version of Darwin's *Origin of Species* and Linnaeus' *Oeconomy of Nature*." *Proceedings*, American Philosophical Society 104 (1960):235–41.

———. "Haeckel, Darwin, and Ecology." *Quarterly Review of Biology* 32 (1957): 138–44.

Steere, William Campbell, editor. *Fifty Years of Botany.* New York: McGraw-Hill, 1958.

Stegmüller, Wolfgang. *The Structure and Dynamics of Theories*, translated from the German by Dr. William Wohlhueter. New York, Heidelberg, Berlin: Springer-Verlag, 1976.

Storr, Richard J. *Harper's University, The Beginnings: A History of the University of Chicago.* Chicago and London: University of Chicago Press, 1966.

Stout, John Elbert. *The Development of High School Curricula in the North Central States from 1860–1918.* Chicago: University of Chicago Press, 1921.

Strangeland, C. E. "Pre-Malthusian Doctrines of Population." *Studies in History, Economics, and Public Law* 21 (1904):1–356.

Struik, Dirk J. *Yankee Science in the Making.* Revised, reprint edition. New York: Collier Books, 1962.

Sutton, S. B. *Charles Sprague Sargent and the Arnold Arboretum.* Cambridge, Mass.: Harvard University Press, 1970.

Tansley, Arthur G. "The Classification of Vegetation and the Concept of Development." *Journal of Ecology* 8 (1920):118–44.

———. "The Development of Vegetation." *Journal of Ecology* 4 (1916):198–204.

———. "Drude's Text-book of Plant Ecology." *Journal of Ecology* 2 (1914):51–53.

———. "The Early History of Modern Plant Ecology in England." *Journal of Ecology* 35 (1947):130–37.

———. *The Psychology and Its Relation to Life.* First edition, 1920. London: George Allen & Unwin, 1922.

———. *Our Heritage of Wild Nature: A Plea for Organized Nature Conservation.* Cambridge [England]: at the University Press, 1945.

———. *Practical Plant Ecology.* Second edition. London: George Allen & Unwin, 1926.

———. "The Problems of Ecology." *The New Phytologist* 3 (1904):191–200.

———. Review of Charles Elton, *Animal Ecology. Journal of Ecology* 16 (1928):163–69.

———. "Succession: The Concept and Its Values." In *Proceedings of the International Congress of Plant Sciences, 1926*, edited by B. M. Duggar. 2 vols. Menasha, Wisc.: George Banta, 1929.

———. "A Universal Classification of Plant-Communities." *Journal of Ecology* 1 (1913):27–42.

———. "The Use and Abuse of Vegetational Concepts and Terms." *Ecology* 16 (1935):284–307.

———. *The Vegetation of the British Isles.* 2 vols. Cambridge, Mass.: Harvard University Press, 1939.

Tobey, Ronald C. *The American Ideology of National Science, 1919—1930.* Pittsburgh: University of Pittsburgh Press, 1971.

———. "How Urbane Is the Urbanite? An Historical Model of the Urban Hierarchy and the Social Motivation of Service Classes." *Historical Methods Newsletter* 7 (1974):259–75.

———. "The Life Cycle of Grassland Ecology, 1895–1955." In *The History of American Ecology*, edited by Frank N. Egerton. New York: Arno Press, 1977.

———. "Theoretical Science and Technology in American Ecology." *Technology and Culture* 17 (1976):718–28.

Tomisch, John. *A Genteel Endeavor: American Culture and Politics in the Gilded Age.* Stanford, Calif.: Stanford University Press, 1971.

Trefethen, James B. *Crusade for Wildlife: Highlights in Conservation Progress.* Harrisburg, Pa.: Winchester Press, 1975.

Trombley, Kenneth E. *The Life and Times of a Happy Liberal: A Biography of Morris Llewellyn Cooke.* New York: Harper, 1954.

True, Alfred Charles. *A History of Agricultural Education in the United States, 1785–1925.* Washington, D.C.: U.S. Government Printing Office, 1929.

Turrill, W. B. *Pioneer Plant Geography: The Phytographical Research of Sir Joseph Dalton Hooker.* The Hague: M. Nijhoff, 1953.

Veysey, Laurence R. *The Emergence of the American University.* Chicago and London: University of Chicago Press, 1965.

Vološinov, V. M. *Freudianism: A Marxist Critique*, translated by I. R. Titunik and edited in collaboration with Neal H. Bruss. New York: Academic Press, 1976.

Wagenitz, Gerhard. "Grisebach, August Heinrich Rudolf." *Dictionary of Scientific Biography*. New York: Charles Scribner's Sons, 1972.

Walker, Helen M. *Studies in the History of Statistical Method*. Baltimore: Williams and Wilkins, 1931.

Wallace, Henry A. *New Frontiers*. New York: Reynal and Hitchcock, 1934.

Walsh, Thomas R. "The American Green of Charles Bessey." *Nebraska History* 53 (1972):35−57.

————. "Charles E. Bessey and the Transformation of the Industrial College." *Nebraska History* 52 (1971):383−409.

Weatherwax, Paul. "Charles Clemon Deam: Hoosier Botanist." *Indiana Magazine of History* 67 (1971):197−267.

Weaver, John E. *Prairie Plants and Their Environment: A Fifty Year Study in the Midwest*. Lincoln: University of Nebraska Press, 1968.

————. *Root Development of Field Crops*. New York: McGraw-Hill, 1926.

————, and Albertson, F. W. *Grasslands of the Great Plains: Their Nature and Use*. Lincoln, Nebr.: Johnsen Publishing Co., 1956.

————, and Bruner, William E. *Root Development of Vegetable Crops*. New York: McGraw-Hill, 1927.

Werskey, Gary. *The Visible College*. London: Allen Lane, 1978.

White, Morton. *Social Thought in America; The Revolt Against Formalism*. Reprint edition. Boston: Beacon Press, 1957.

Whittaker, Robert H. "Classification of Natural Communities." *Botanical Gazette* 28 (1962):1−239.

————. "Recent Evolution of Ecological Concepts in Relation to the Eastern Forests of North America." In *Fifty Years of Botany*, edited by William Campbell Steere. New York: McGraw-Hill, 1958.

Whorton, James. *Before Silent Spring: Pesticides and Public Health in Pre-DDT America*. Princeton: Princeton University Press, 1974.

Wigdor, David. *Roscoe Pound, Philosopher of Law*. Westport, Conn.: Greenwood Press, 1974.

Wiley, Farida A., ed. *John Burrough's America*. Old Greenwich, Conn.: Devin-Adair Co., n.d.

Williams, Raymond. *The Country and the City*. Reprint edition. St. Albans [England]: Paladin, 1975.

Wilson, Leonard G. *Charles Lyell; the Years to 1841: The Revolution in Geology*. New Haven and London: Yale University Press, 1972.

Wohl, R. Richard. "The 'Country Boy' Myth and Its Place in American Urban Culture: The Nineteenth-Century Contribution," edited by Moses Rischin. *Perspectives in American History* 3 (1969):77−156.

Worster, Donald. *Nature's Economy: The Roots of Ecology*. San Francisco: Sierra Club Books, 1977.

Zaunick, Rudolph. "Drude, Carl Georg Oscar." *Neue deutschen Biographie* 4 (1957).

INDEX

also Bessey, Charles; Goethe, J. W.; Idealism; Ideal Type

Metamorphosis of Plants, by Goethe, 43, 59. *See also* Goethe, J. W.

Meter-plot method, 70; originated to determine abundance, 59–60. *See also* Abundance; Clements, Frederic; Impressionism; Pound, Roscoe; Quadrat

Methodology, quantitative research and analysis for case study of grassland ecology, 223–250

Michigan Agricultural College, 10, 12

Michigan State College, 132

Microparadigm: anomalies of, 115; and apprenticeship in learning, 114; chronological analysis of cumulative bibliography of, 117; as cognitive orientation to scientific research, 37, 113; and collaborative support network, 115, 117, 118; and competition between candidates, 108–109; cumulative publications on, 115–117; decline in theories of, 115; gestalt-switch to, 114; as intellectual innovation, 111; and Kuhn's philosophy of science, revision of, 110–119; as methodological innovation, 113; normal science in, 114–115, 117; and parallelism between intellectual and social growth, 110; and plant succession as basis of microparadigm for plant ecology, 79; sociology of, 110. *See also* Clementsian microparadigm, Crane, Diana; Kuhn, Thomas

Microscope, as basis of New Botany, 10, 11. *See also* New Botany

Microscopical Society, 12

Midwest, 2, 5, 11, 27, 53, 164, 217

Miller, Howard, 34

Millikan, Robert, 125

Milton, Ohio, 10

Mississippi River, 211

Missouri Botanical Garden, 34

Missouri prairies, Drude's regionalization of, 64. *See also* Clements, Frederic; Drude, Oscar; *Handbuch*; Pound, Roscoe; Prairies

Missouri River, 53, 79

Mitchell, Maria, 33

Mitchell, S. Weir, 33

Montana State College, 132, 220

Morgan, C. Lloyd, 161, 167, 174, 175

Morgan, Thomas Hunt, 14

Muir, John, 1, 24

Mullkay, Michael, 112, 113

Naegeli, Karl W. von, 102

National Conservation Congress of 1912, 14

National Education Association, 13, 30, 38

National Environmental Policy Act, 1

National Science Foundation, 222

Natural history: in genteel tradition, 32; in idealism, 33

Naturalism, ecology opposed to, 72–73

Natural selection. *See* Survival of fittest

Nature appreciation, and antiurban sentiment, 26. *See also* Antiurbanism

Nature Conservancy Movement, 187, 188

Nature study movement, 24, 38. *See also* Bessey, Charles; Coulter, John

Naturphilosophie, 44, 88, 89, 90; in Sachs's *Lehrbuch*, 41–45

Nebraska, 4, 9, 13, 19, 20, 27, 53, 57, 58, 60, 61, 62, 68, 77, 84, 88, 111, 191, 193, 201; Horticultural Society, 13, 14; Park and Forestry Association, 14

Nebraska Wesleyan, 132

Neolamarckism, 85–87, 182. *See also* Lamarck

Neoplatonism, 32

New Botany, 11, 14, 21, 24, 29, 30, 31, 35, 37, 102. *See also* Bessey, Charles; Coulter, John; Microscope

New Deal, 8, 209, 210, 212, 214

Designer: Kitty Maryatt
Compositor: Trend Western
Printer: Thomson-Shore
Binder: Thomson-Shore
Text: 11/13 Janson